PASS IN REVIEW

OSPREY
PUBLISHING

PASS IN REVIEW

AN ILLUSTRATED HISTORY OF
WEST POINT CADETS
1794–PRESENT

Clyde W. Cocke

Class of 1977, United States Military Academy

Photos by Eilene Harkless Moore

First published in Great Britain in 2012 by Osprey Publishing, Midland House, West Way, Botley, Oxford, OX2 0PH, UK
44-02 23rd Street, Suite 219, Long Island City, NY 11101, USA

E-mail: info@ospreypublishing.com

The views expressed by the parties named herein do not necessarily reflect the official views or policies of the United States Military Academy.

Unless otherwise noted, photos are by Eilene Harkless Moore.

Clyde Cocke has asserted his right under the Copyright, Designs and Patents Act, 1988, to be identified as the author of this work.

A CIP catalogue record for this book is available from the British Library.

ISBN: 978 1 84908 558 8
Page layout by Tom Heffron, Hudson, Wisconsin, USA
Index by Sharon Redmayne
Typeset in Adobe Caslon Pro
Originated by United Graphics Pte., Singapore
Printed in China through Bookbuilders Ltd.

12 13 14 15 16   10 9 8 7 6 5 4 3 2 1

Osprey Publishing is supporting the Woodland Trust, the UK's leading woodland conservation charity, by funding the dedication of trees.

www.ospreypublishing.com

CONTENTS

FOREWORD

As I heard the command "Pass in Review" and stood with my fellow classmates of 1951 during our sixtieth reunion, I recalled my own West Point experience. The cadets marched forward onto the Plain while my own memories marched backward into time.

I called to mind the challenges of Beast Barracks starting on July 1, 1947. I was only seventeen years old when I took the oath as a cadet. Marching and drilling in formation on the Plain was just one of the many ways I learned discipline while at West Point. I remembered the parade in the summer dress uniform of India white during July 1948 as our class represented West Point at the funeral of General John Pershing, Class of 1886. I reminisced about my own marching in the ranks of Cadet Company G-1. Only eighteen of twenty-seven plebes who joined G-1 survived the rigors of West Point. One of these G-1 cadets, Roscoe Robinson Jr., would become the first African-American four-star Army general. We proudly marched with the entire Corps of Cadets for the inauguration parade of President Harry Truman in January 1949.

I learned the value of teamwork while on the track team. I reflected on the fate of my track teammate, Ed White, Class of 1952. We both donned Air Force blue after graduating from West Point. I fought in the Korean War along with most of my classmates and flew sixty-six combat missions. I then joined Ed White in Germany as a fighter pilot. We again served together as astronauts. Tragically, Ed White perished

Cadet Buzz Aldrin, shown here in 1951. (Author's Collection. Photo by Eilene Harkless Moore)

in a fire on January 27, 1967, while conducting training in the Apollo I command module. He is the only United States Military Academy graduate who made the ultimate sacrifice for space exploration. His final resting place is at the West Point cemetery.

As a cadet, I experienced a variety of academic classes and social activities. The four-year engineering curriculum prepared me to later earn a doctorate in astronautics on manned space rendezvous at the Massachusetts Institute of Technology. The dancing classes I took with my classmates and weekly cadet dances called "hops" would serve me well when I performed on "Dancing with the Stars" on national television.

Little did I know, as I passed in review during my last cadet parade, that I would go from marching on the Plain to walking on the Moon.

—Buzz Aldrin
United States Military Academy, West Point
Class of 1951

Opposite: Buzz Aldrin standing on the moon in July 1969. (Photo courtesy of NASA)

Buzz Aldrin at the sixtieth reunion of the Class of 1951.

INTRODUCTION

To the casual observer, a cadet parade (in military parlance called a "Review") at the United States Military Academy, West Point is just fleeting entertainment with stirring martial music and marching cadets resplendent in their full dress uniforms. To the informed viewer, however, every aspect of this event is imbued with a deeper historical meaning.

The history starts with the parade ground itself, known as the "Plain." The Plain has born silent witness for more than two hundred years to ruly ranks of first blue- then gray-coated cadets. The Plain has felt the trod of thousands of feet. It has heard the shouted commands from first captains such as Douglas MacArthur, adjutants such as George Patton, and Academy superintendents such as Robert E. Lee. The Plain has seen countless salutes exchanged between the Corps of Cadets and U.S. presidents such as Eisenhower and generals such as Schwarzkopf.

Though spectators of the Reviews have changed over the years, little else has on the Plain, where tradition holds fast. First on and last off the Plain is the United States Military Academy Band. This unit of active-duty soldiers predates the founding of the Academy by many years. Fifers and drummers were part of the American Revolutionary War companies that garrisoned West Point beginning in 1778. In 1817, the musicians were named the West Point Band. The band is authorized to design and wear its own unique uniform. The present-day full dress uniform incorporates items from both the Army blue and the cadet full-dress uniforms.

Statues of George Patton (left foreground) and Dwight Eisenhower (left middle ground) silently keep watch on the Plain.

The band marches onto the Plain before the start of a review.

Cadets begin their march onto the Plain from the various cadet barracks by deploying in company formations through passages called "sally ports." As of 2011, thirty-two companies were organized in a brigade of four regiments. Each regiment consists of two battalions with four companies per battalion. Each year, roughly a dozen parades are conducted; five involve the entire brigade. Most home football game parades, held on Saturdays during the autumn, consist of just two regiments.

Companies line up by use of a right guide who posts to the appropriately marked small, gold-and-gray pennant. This allows each company to line up precisely on the Plain by battalion and regiment. The color guard marches in the center of the formation with the American, Army, and Corps colors along with two escorts.

Cadet companies march out of a sally port onto the Plain.

An Air Force Cadet (wearing blue coat) acts as the right guide to mark the spot for the rest of the company to form up on.

Once formed up on the Plain, the reviewing party moves forward to review the parade. Typically, the superintendent, the commandant, the dean, and honored guests compose this group. The cadet first captain, who commands the Corps of Cadets, and the brigade staff lead the cadets in saluting the reviewing party.

After appropriate honors and salutes have been rendered to the reviewing party, the order is given to "Pass in Review," and the march-

Companies lining up on the Plain with colors on right.

Companies deployed on the Plain with the Cadet Chapel in the background (left).

by begins. The first captain and brigade staff initiate "eyes right" as they march in front of the reviewing party. The band follows and assumes a standing formation to play appropriate military tunes as the rest of the Corps parades by the reviewing party. Each company, as part of their eyes right, lowers their guidon parallel to the ground as they pass the reviewing party.

Only the color guard does not dip all its colors to the reviewing party. The United States flag is never lowered to anyone during a parade at any time. The Army flag is dipped only to a person of authority over the Army, such as the President of the United States. The Corps of Cadet colors are lowered because the superintendent has command over the Corps.

The two Cadet Color Sergeants with rifles do not have bayonets on their rifles in order to prevent the wind from possibly tearing the flags on the sharp-tipped bayonets. All other cadets march with fixed bayonets at "right shoulder arms."

The passing of the color guard signals that half the companies have passed in review because the colors are always centered in the

The first captain (right) leads the bridge staff as they initiate "eyes right" to salute the reviewing party.

procession. The trail regimental staff follows the colors along with the remaining eight companies. The lead regimental staff previously marched in front of the the first eight companies.

Exchange cadets from the Naval and Air Force Academy march in the parade along with allied cadets from various countries, such as the Royal Military College of Canada. Their different uniforms, like the red-coated Canadians, stand out in the sea of gray.

Last off the Plain is the band. The director of the band is the only officer marching with the musicians. He carries a sword and wears the standard Army Blue service cap. The band returns to Egner Hall, their

The drum major (right) and band director (left front) perform "eyes right" for the band.

The band plays appropriate martial music for the marching cadets.

A company marches by
the reviewing party at
"eyes right."

Cadet colors are
lowered to the
reviewing party.

A midshipman from the Naval Academy, in black uniform and white cap (right front), marches as part of the Academy exchange program.

A red-coated cadet from the Royal Military College of Canada stands out in the sea of gray.

The band marches in front of the superintendent's quarters en route to its own.

The band is the last to march off the Plain with the band director (left front) carrying a cadet saber.

rehearsal facility and headquarters, by way of Quarters 100, the superintendent's house.

The cadets march back to their respective barracks and pass by a statue of George Washington mounted astride his horse. He is honored both as the first Commander of the Army and the first President of the United States. In one of his last letters, written in 1799, Washington advocated for the creation of a military academy. Thomas Jefferson signed into law an act of Congress that established a military academy at West Point on March 16, 1802, henceforth celebrated as Founders Day.

Following the parade, the Plain remains, again, silent with the newest memories of where the Corps has trod. Only time will tell if a future president, general, astronaut, or Medal of Honor winner has passed in review.

The Plain again stands silent, waiting for the next Pass in Review.

CHAPTER 1

The First Cadets

The Corps of Artillerists and Engineers
1794–1801

West Point cadets were first authorized by an act of Congress on May 9, 1794. The act established the Corps of Artillerists and Engineers with 992 soldiers. A new rank of "cadet" was created for the American Army, and precedents were set which survive to this day. The cadets were part of the regular Army and subject to discipline under the articles of war, now the Uniform Code of Military Justice. Cadets were paid the same as a sergeant, at the time $6 a month. Congress would tinker with the pay rate of cadets during the next two hundred years, but they are still paid as part of the active-duty Army.

Cadets also received the clothing allowance of a sergeant. Beginning in 1802, they had to provide their own uniforms. Cadets and officers still pay for their own uniforms, while the enlisted ranks receive a uniform allowance.

Thirty-two cadets of the Corps of Artillerists and Engineers were authorized under the Act. Two cadets were allocated to each of the sixteen companies, which were organized into four companies per battalion. The entire Corps was composed of four battalions and a staff. The few surviving records indicate that only a handful of cadets were part of this Corps: just three cadets in 1795 as well as in 1796.

Although this act did not specify West Point as the location for the Corps of Artillerists and Engineers, the previous history of West Point, dating back to the American Revolutionary War, ensured the Corps would be located in New York State at West Point. The strategic location

This statue of Polish engineer Tadeusz Kosciuszko overlooks the Hudson River. Kosciuszko was West Point's first chief engineer.

1

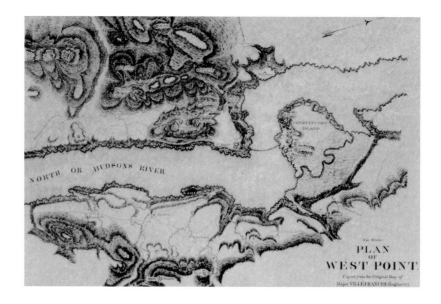

of West Point on the Hudson River, located fifty miles north of New York City, made it a vital military spot. The area juts out into the river as the "west point" that forces ships to sail very slowly to maneuver around the S-shaped bend. General George Washington recognized the importance of defending the Hudson River and advocated fortifying key terrain on the river. If the British controlled the waterway, they could isolate New England from the rest of the colonies.

On January 20, 1778, American soldiers first occupied West Point, and it remains to this day the oldest continuously occupied post of the U.S. Army. Colonel Tadeusz Kosciuszko, a Polish engineer trained in France, arrived in March 1778 as the chief engineer for West Point. He built a series of fortifications to form Fortress West Point. A barrier called the Great Chain was laid across the river between West Point and Constitution Island, located on the east bank of the river. General Washington made the first of many visits to West Point in the summer of 1778.

Just two years later, West Point survived a treasonous plot by its commander, Major General Benedict Arnold, in September 1780. He conspired with the British to turn over West Point to them. Fortunately for the American cause, Arnold's British contact, Major John André, was captured and the plot was discovered. Arnold escaped to a British warship, but his name is forever linked to treason and betrayal.

After the failed treason by Arnold, West Point saw no further action during the Revolutionary War. The signing of a peace treaty with

Great Britain in 1783 resulted in the American Army being reduced to just eighty soldiers and a few officers by June 1784. A fifty-five-man detachment of artillery guarded West Point and the arms and equipment stored therein.

In November 1793, President Washington's cabinet considered establishing a military academy. While Secretary of the Treasury Alexander Hamilton and Secretary of War Henry Knox favored forming a military academy, Secretary of State Thomas Jefferson opposed it as unconstitutional. Washington wanted to avoid further disagreements among his already divided cabinet, so he did not press the issue with Congress. Congress responded to Washington's concerns about officer education by establishing cadets within the new Corps.

A detachment of artillery garrisoned West Point, and the expanded Corps of Artillerists and Engineers was initially stationed at West Point. Washington had supported a military academy since the Revolutionary War, but he was unable to convince Congress to create one.

In 1798, Congress enlarged the Army in response to the threat of war with France. The Corps of Artillerists and Engineers was expanded to two regiments and the number of cadets increased to sixty-four.

Links from the Great Chain on display at Trophy Point looking north toward the Hudson River.

3

Cadet pay was specified as $10 per month. Four "teachers of the arts and sciences" were authorized to instruct the artillerists and engineers. President John Adams, also an ardent supporter of a military academy since the Revolutionary War, wanted to appoint Americans as these instructors. However, Adams did not fill any of the teaching positions before he left office in March 1801 after failing to win a second term. He appointed Jonathan Williams as an officer in the Corps of Artillerists and Engineers and appointed five more cadets into the two regiments of Artillerists and Engineers. Adams laid the foundation for Jefferson to establish a school at West Point in 1801.

Jefferson started a school at West Point by employing existing laws to appoint instructors, cadets, and an officer to act as superintendent. Jonathan Williams was appointed Inspector of Fortifications by Jefferson and ordered to West Point to direct the school. George

Captains of infantry (left foreground) and artillery (right foreground) confer at West Point with soldiers (background), 1783–96. (West Point Museum, Photo by Eilene Harkless Moore)

Baron, an Englishman, was employed as the teacher of mathematics. Fourteen cadets are known to have attended the school at West Point beginning in autumn 1801. Baron used a blackboard for instruction. This was the first use of a blackboard in America, and it became a prominent feature in American classrooms for decades to come. Cadets would take "boards" as part of their mathematics instruction and present their solutions to problems on a daily basis into the late twentieth century. The now-titled Department of Mathematical Sciences celebrated its bicentennial on September 21, 2001, the anniversary of Baron's first class.

Class started at 8:00 a.m. and lasted until noon. Military drill and field sports were held in the afternoon. Instruction was conducted in a wood-framed, two-story building called the "Academy." There were two rooms on the ground floor, but the small number of cadets required the use of only one. The three rooms on the second floor were used as officers' quarters. Cadets sat on low benches and used the blackboard for recitation.

Four of these cadets were commissioned into the two regiments of Artillerists and Engineers in 1801; one was discharged. The others were present on March 16, 1802, when the Military Academy was founded by an act of Congress as part of the newly established Corps of Engineers, and seven would eventually graduate from the Military Academy.

Joseph Swift was made a cadet via a warrant granted by President John Adams on May 12, 1800. He was attached to a company of the artillerists and engineers located at Fort Wolcott located near Newport, Rhode Island. Swift, now recognized as the first graduate of the Military Academy, actually almost failed to graduate. He reported to West Point in October 1801 to attend the newly opened school with Baron as the instructor. Swift was welcomed into an artillery mess. Baron wanted Swift to dine with the other cadets and sent a servant to

George Washington's statue sits in front of Washington Hall overlooking the Plain.

instruct Swift to leave the artillery mess. Swift refused to be ordered by a mere servant and Baron then confronted Swift outside the artillery quarters. Baron called Swift "a mutinous young rascal." Swift sprang at Baron to hit him. Baron retreated to his quarters and shut the door in Swift's face. They exchanged coarse insults, and Swift was charged with being disrespectful to a superior. It was an inauspicious start to the cadet's career. However, before action was taken against him, Baron was charged by the post commandant with conduct unbecoming. Both were convicted, but Baron was dismissed while Swift was reprimanded and allowed to graduate the next year. In his memoirs Swift wrote, "Mr. Baron was of a rude manner but he was an able teacher."

In addition to being the first graduate of West Point, Swift left yet another lasting legacy at the academy. A small garden planted by Colonel Kosciuszko during the 1770s was located on the eastern edged of the Plain. Swift repaired it while a cadet during the summer of 1802, and the garden still exists.

But not everything lasts forever, and cadet barracks are among those things that have come and gone over the years. The first cadet barracks were located on the northeast side of the Plain. This building, called Long Barracks because at 240 feet it was the longest building at West Point, was made of wood and burned down in 1827. At first, it

This painting from the 1790s depicts soldiers formed up in front of Long Barracks on the Plain. (West Point Museum, Photo by Eilene Harkless Moore)

was used by the enlisted soldiers of the Artillery Company stationed at West Point. There was no requirement for cadets to billet or mess together. Cadets would lodge with families of the officers residing on the grounds of West Point. Later, when cadets were required to live together, Long Barracks became the first cadet barracks.

Cadet uniforms date back even farther than cadet barracks. Uniforms for cadets were first mentioned in 1795 in the clothing book of the First Regiment of Artillerist and Engineers. Three cadets are charged with swords, sword belts, sergeants' helmets, plumes, cockades, and sergeant's coats and shirts for their personal use. The hats were round with a brim three inches wide with a strip of bearskin seven inches high across the crown. A black cockade, eagle, and red plume completed the hat.

In 1799, the uniform of a cadet of artillery was prescribed as a blue coat with white buttons and red facings, white underclothes and cocked hat. Officers' coats reached to the knees, and the artillery coats were lined with red. Cadets were distinguished by a silver strap on the right shoulder. They also wore swords, as did the officers. A black cockade, symbol of a soldier since the Revolutionary War, was worn on the hat. A small white eagle was centered in the cockade, also dating from the Revolutionary War. Cadets wore red plumes and on the left shoulder a gold strap with fringe.

Kosciuszko's garden, first planted in the 1770s, still exists at West Point.

A cadet (center), 1801, wearing gold strap on his shoulder and carrying a sword along with soldiers of the Artillerists and Engineers at West Point. (West Point Museum, Photo by Eilene Harkless Moore)

The first cadet uniforms of 1801 were modified from the earlier one but were still those of the Corps of Artillerists and Engineers. Cadets dressed similarly to the junior officers and wore a sword, to this day the mark of an officer, suspended from a white leather shoulder belt. The uniform consisted of a dark blue cutaway coat with red facing, brass buttons, white or blue waistcoat and trousers, black leather boots, and a cocked hat with a black cockade and red-feathered plume. Cadet rank was indicated by a blue shoulder strap edged with gold lace instead of an officer's epaulette. A sword knot, also a symbol of an officer, was not allowed until the cadet had been commissioned as an officer.

This school, formed by President Thomas Jefferson in 1801 during his first year in office, was the product of several attempts by the previous two Presidents—Washington and Adams—to create a military academy. The Cadets of the Corps of Artillerists and Engineers would constitute the Cadets of the Military Academy. Williams, head of the school, would be in charge of the Academy.

CHAPTER 2

Before Cadet Gray

1802–14

The Military Peace Establishment Act of March 16, 1802, created the United States Military Academy as part of the Corps of Engineers, stating "that said corps, when so organized, shall be stationed at West Point in the state of New York, and shall constitute a military academy," thus intertwining the Academy and the Corps of Engineers. This act also directed the "principal engineer" to be the superintendent of the Academy, further joining the two organizations together—with unintended consequences for the prospects of the future Academy.

Establishment of the Academy resulted in little immediate change at West Point. While today graduates around the world celebrate March 16 as "Founders Day," there was minimal impact on the cadets at West Point with the formal founding of the Academy. The cadets already present at West Point from the 1801 school remained. Seven cadets from this school would graduate from the newly formed Academy. Four cadets from the 1801 school became the first four graduates of the Academy, with two in the Class of 1802 and two of the three in the Class of 1803. The other three later graduated.

Several factors conspired against the growth and even the survival of the fledgling Academy. The status of the superintendent was the most vexing. Jonathan Williams, although a Federalist, was retained by Jefferson to fill both positions. Wearing two hats, as both head of the new Academy and of chief engineer, resulted in frequent absences from

This statue of Thomas Jefferson is located in Jefferson Hall, home of the new Cadet Library, which opened in 2008.

Jonathan Williams, the Academy's first superintendent. (West Point Museum, Photo by Eilene Harkless Moore)

West Point to fulfill the engineering duties by inspecting fortifications. Williams, a Harvard graduate, was a grandnephew of Benjamin Franklin and had served as Franklin's secretary in France. Williams knew Jefferson, having first met him in France while Franklin was serving as ambassador. Jefferson believed Williams to be a man of science like his kinsman Franklin. Jefferson also considered Williams to be a moderate Federalist not associated with the hard-core faction of the party aligned with Alexander Hamilton.

Jefferson used his appointment powers to select the officers of the newly formed Corps of Engineers. Although Williams and some other perceived moderate Federalists were retained, others were cashiered. Jefferson, the first President from the then-called Republican Party, now the present-day Democrats, wanted to place as many Republicans into the federal government as possible.

Only twenty positions were authorized within the Corps of Engineers. Initially there were ten officers and ten cadets. However, at the discretion of the President, up to sixteen officers could be appointed, leaving only four cadet billets available at West Point. No enlisted soldiers were included as part of this Corps. The reduced Artillery Regiment still retained forty cadet positions. The remaining two infantry regiments were not authorized cadets within their companies.

There were no qualifications required for cadets to attend the new Academy. In fact, a lack of standards for admission was a factor in limiting the ability of the Academy to advance beyond an "elementary mathematics school." No set time to report for instruction was established, nor was an age limit established. For example, John Lilly, eleven years old, was appointed out of sympathy in December 1801. His father, the military storekeeper at West Point, died in September 1801. Jefferson then appointed Lilly, but he did not graduate and resigned in 1805.

Engineers were prohibited from commanding other soldiers based on regulations dating back to the Revolutionary War, when foreigners were appointed as engineers. In September 1802, Williams, now a lieutenant colonel, found himself in an awkward position. He had to requisition supplies through Captain George Izard, commander of the Artillery Company. Izard could and did disapprove some requested items, which greatly bothered Williams. Secretary of War Henry Dearborn initially ignored Williams's request to change the regulations and make engineers part of the regular Army. Williams resigned over this issue in June 1803, after a final request to Dearborn was rejected.

There was an upside for cadets of the Corps of Engineers. The act authorized their pay at $16 a month versus the $10 monthly pay of cadets from the Artillery Regiment. Swift and Simon Levy, the second graduate of the Academy, both transferred to the Corps of Engineers as the first cadets of this corps in July 1802. Levy was twenty-five years old and was appointed a cadet because of his valorous service in the Indian Wars of 1794 in northwest Ohio. He also attended the 1801 school, arriving at West Point in March of that year. The first Jewish graduate of the Academy, he resigned from the Army due to illness and died in 1807.

The faculty was formed in spring 1802 with Captain William Barron of the Corps of Engineers. Like Williams, he also was a graduate of Harvard, where he had been a tutor. A Yale graduate, Jared Mansfield, was appointed as a captain in the new Corps of Engineers. Mansfield had taught mathematics at Yale, written a book about mathematics, and—better yet for Jefferson—was a fellow Republican.

July 4, 1802, marked the formal opening of the United States Military Academy at West Point. Just three men comprised the staff and faculty: Williams as superintendent and Barron and Mansfield as teachers. All three taught the ten cadets, with Barron and Mansfield teaching mathematics and Williams instructing on surveying and fortifications. Classes started at 9:00 a.m. and ended at 2:00 p.m. Starting a tradition that lasted into the twenty-first century, Saturday classes also were held, with only Sunday as a free

This cadet, from 1802, wears white pantaloons for summer and carries a sword. (West Point Museum, Photo by Eilene Harkless Moore)

day. On four days during the week, field exercises were held in the afternoons from 4:00 p.m. until sunset. To this day, cadets participate in intramural athletics or military training in the afternoon.

Swift and Levy were given oral examinations by the faculty in October 1802. They were tested on mathematics, natural philosophy (physics), artillery, and field fortification. Upon successful completion of the examinations, they were promoted to second lieutenants in the Corps of Engineers on October 12, 1802.

This portrait portrays Joseph Swift, the first graduate of the Academy, as a junior officer. (West Point Museum, Photo by Eilene Harkless Moore)

One practice that did not last, to the regret of cadets, was to take a winter leave from December 1 to March 15. Williams convinced Secretary of War Dearborn, to whom he reported, that the snows of West Point and the inadequate buildings prevented the conduct of classes. The arrival of Sylvanus Thayer as superintendent in 1817 ended this extended winter leave, subjecting cadets to the "gloom period" of long nights, gray skies, deep snows, and icy winds blowing off the Hudson River.

Congress expanded the faculty and addressed the problem of no enlisted soldiers for the Corps of Engineers by an act passed in February 1803. Nineteen soldiers were added to the corps, and new faculty were authorized for the Academy. Two new positions, a teacher of French language and a teacher of drawing, were authorized by Congress. Both were attached to the Corps of Engineers with their pay set as that of a captain. Williams originally filled both positions with a Frenchman, Francis de Masson. In 1808, Christian Zoeller was appointed as the drawing teacher; de Masson continued as the French teacher until 1812.

The abrupt resignation of Williams in June 1803 exposed the problem with the superintendent also being the chief engineer of the Army. Major Decius Wadsworth was now the principal engineer and also became superintendent, but he was supervising harbor fortifications along the Atlantic coast. He was only at West Point sporadically starting in late spring 1804. But his tenure was not long. He took a leave of absence because of ill health in autumn 1804 and resigned in early 1805.

Captain Barron became the acting superintendent because he was the senior engineer at West Point. Like Williams, Barron had no military background when commissioned into the Army. He was an able mathematics teacher but a poor administrator.

Williams was convinced by General James Wilkinson, the commanding General of the Army, to return to West Point in April 1805. He arrived in time to deal with new challenges. Captain Mansfield had left in 1803 to become the surveyor general of the Northwest Territory, leaving just Barron and de Masson as instructors.

When Williams returned in 1805, cadets were billeted in the Long Barracks along with the enlisted soldiers of the Corps of Engineers and those of the Artillery detachment. Unlike present-day barracks, which are furnished with government-issued furniture, no items were supplied to the cadets in the early days of West Point. They slept on wooden pallets on the floor, warmed by a fireplace. Buckets of water were kept both for personal use and in case the fire in the wooden barracks needed to be extinguished quickly. Cadets scrounged boxes and other items to use as tables and chairs.

Upon his return to West Point, Williams instituted a more rigorous schedule. Infantry drill took place from 5:00 a.m. to 6:00 a.m., followed by classroom instruction in mathematics from 8:00 a.m. to 11:00 a.m. French and drawing were taught on alternate days from 11:00 a.m. to 1:00 p.m., followed by a study period in the barracks from 2:00 p.m. to 4:00 p.m. Practical gunnery, artillery drill, study of field fortifications, or surveying was conducted from after 4:00 p.m. until sunset. Lieutenant Alexander Macomb was the adjutant as well as the tactics instructor. He used a drill book first written by Baron Von Steuben for use by American soldiers during the Revolutionary War. Barron continued to teach mathematics, while Masson taught French and drawing to the twenty-four cadets present in April 1805.

Williams established the first cadet mess in August 1805. The inefficient separate messes were consuming the cadets' pay, leaving them with no money left for uniforms. Most cadets took their rations in cash, at twenty-seven cents per day. The cadets often were in debt because pay was irregular. They used their money to buy food for hired cooks to prepare, and a mess steward was hired to run the mess. The mess closed in early 1807 due to financial and administrative problems as the mess lost money. The mess steward was fired and the mess disbanded. Cadets returned to taking their meals with local families until a second cadet mess opened in 1813.

Cadet uniforms were another challenge Williams attempted to solve upon his return. There was no single cadet uniform. Those cadets in the Artillery Regiment wore the uniform of an Artillery officer, while those in the Corps of Engineers wore the uniform of the corps. Both uniforms featured a high-collar, dark-blue coat with gold buttons. Engineer cadets wore black facings on the coat while Artillery cadets continued with their traditional red facings. Another practice that exists to this day was wearing different winter and summer uniforms. The winter uniform consisted of blue woolen pants called "pantaloons" with a blue vest. In summer, white linen pants and vests were worn. Black, cocked hats were worn year-round. Cadets decided if they wanted to wear the half boot of an officer or the cheaper enlisted black gaiters.

Swords also were optional as some cadets could not afford to buy a sword, the symbol of an officer.

Under Barron's lax command, cadets wore any uniform they pleased. Some reported in militia uniforms they had bought before reporting to West Point. One cadet wore a general's coat. Many uniforms were tattered and worn because cadets were spending all their money on food prior to the cadet mess being established. Williams asked Dearborn to issue regulations for a common cadet uniform, regardless of branch. Dearborn did not issue regulations for a new uniform, but instead directed cadets to buy uniforms of their branch. A prescribed cadet uniform would not be established until 1810.

An engineer soldier 1803–05 (left) shown wearing artillery hat with red plume; engineer soldiers 1806–11 (center) wearing round hats with black plume; and drummer 1811 (right) in red coat. (H. Charles McBarron, Company of Military Historians, Photo by Eilene Harkless Moore)

Williams also asked the secretary of war to authorize a clothing allowance for the cadets. However, the parsimonious Dearborn not only refused an allowance, but, to save even more money, he appointed cadets only to the Artillery Regiment starting in 1805. This saved the extra six dollars a month that the Corps of Engineer cadets had been receiving since 1802 and also forced all cadets to wear the Artillery uniform. Williams borrowed enlisted Artillery uniforms from Colonel Henry Burbeck, the commander of the Artillery Regiment, to make for a more consistent appearance among cadets.

Williams continued to advocate for a distinctive cadet uniform, and one finally was authorized on April 30, 1810. The cadet-issued new style of dress was based on Jefferson's recent uniforms for the Army. Jefferson wanted a "Republican" style of dress based on the new French fashion of simpler, more egalitarian dress style established in the wake of the fall of the French monarchy and the French Revolution.

The American Revolutionary War-type uniform with powdered hair, cocked hats, cutaway coats, buckled shoes, and knee breeches would soon disappear. Jefferson personally introduced long trousers, then called pantaloons, and buttoned shoes. Cadets traded in their cocked hats for a "round hat" similar to the present-day high hat or top hat. Long, powdered hair was discarded for shorter, cropped hair, which ensured that hair length would forever remain a topic for debate among cadets at West Point.

A dark blue coat, cut across the waist, with a black braid that buttoned down the front replaced the previous cutaway. For work or everyday use, a version called a "coatee" was worn with shorter tails than the full dress coat with longer tails. Tightly worn pantaloons reached down to the ankles.

When "under arms" with musket and bayonet, cadets wore white cross belts fastened in the front by a brass plate. One belt held a black leather cartridge box and the other a bayonet scabbard. White belts have been worn ever since on parade. Swords were still part of a cadet's attire, but all cadets did not own one because uniform items were purchased by a cadet and, as mentioned, not everyone could afford a sword.

Threat of war once more with Britain in 1807 caused Congress to increase the size of the Army and, hence, West Point in 1808. Williams called for more cadets and three permanent professors. Congress ignored most of his requests, opting only to add cadets to the newly formed regiments of the Army. In total, 156 new cadets were authorized in all branches including infantry, cavalry, and the newly created light artillery. The maximum number of cadets authorized at West Point was now 206. Dearborn refused to appoint cadets from the infantry and cavalry regiments.

West Point continued to have problems with the election of James Madison in 1808. The actions of the new secretary of war, William Eustis, almost led to the end of West Point. Whether through incompetence or malice toward West Point, he appointed only four cadets in 1809, only two in 1810, and none in 1811. Eustis also ordered all cadets to their regiments and the engineer officers to build harbor fortifications to prepare for war with Britain. This resulted in no cadets graduating in 1810. By June 1812, there were no cadets or instructors

From 1805, this cadet is shown in summer uniform wearing half boots with an enlisted engineer soldier (right) wearing blue coat and black velvet collar and cuffs. (U.S. Army print, Photo by Eilene Harkless Moore)

at West Point, and the school lay dormant for almost a year. Only the impending war with Britain in 1812 caused Congress to finally address Williams's proposals.

A seminal event in the history of the Academy came with the passage by Congress on April 29, 1812, of an act second only in importance to the founding in 1802. This act established the Corps of Cadets as a separate entity with two hundred fifty cadets. They were no longer appointed as part of a regiment. Instead, cadets were to "be attached at the discretion of the President ... as students to the military academy."

Congress left the cadet spaces within the Corps of Engineers so there could be four to ten more cadets in addition to the two hundred fifty authorized by the act of 1812. The best news for the cadets was a pay raise as Congress made the engineer cadet pay of $16 a month standard for all cadets.

This bill also provided the legal foundation for the organization of the

Cadets in 1812 shown wearing the first unique cadet uniform of blue coats and round hats with black cockade in winter dress with gray pantaloons.
(West Point Museum, Photo by Eilene Harkless Moore)

Academy, which continues to this day. The academic framework was established with permanent faculty, separate from the Corps of Engineers. Professors of natural and experimental philosophy (what we would call physics today), mathematics, and the art of engineering were authorized in addition to the previous teachers of drawing and French. Assistant professors also were allowed for each of the three professors. These academic positions could be filled either by civilians or engineer officers and were paid the same as officers. The academic staff would determine standards for graduation and cadets would receive a degree. Cadets were commissioned by the President into any corps of the Army.

The new law also required of cadets "that they shall be arranged into companies of noncommissioned officers and privates," and cadets today continue to live in the barracks within companies. Engineer officers were directed to provide military instruction to the cadet company, which today is done by a tactical officer for each company, who is the legal commander of the cadet company. Four musicians were authorized for each cadet company. Cadets were to "be encamped at least three months of each year," which started the practice of military training in the summer, a practice that continues to this day.

In addition, the law allocated $25,000 for new buildings at West Point. A library and "all necessary implements" were specified in the act.

Williams had at last achieved his goals for the Academy. Upon declaration of War with Britain in June 1812, Eustis told Williams that he would command the New York City harbor fortifications. Eustis then changed his mind, and Williams resigned in July 1812. After his resignation, engineer officers were allowed to command forts and other Army units.

Williams's main accomplishment as superintendent was to maintain the Academy's existence. The organizational flaw of also being the chief engineer kept him away from West Point for extended periods of time and from performing his duties as superintendent.

Following Williams's departure, Swift became superintendent of West Point in July 1812. Swift, the first graduate of the Academy, also had been the senior engineer. His duties as chief engineer in fighting the British during the War of 1812 limited his time at West Point.

In his new position, Swift planned to implement a new four-year academic program and to broaden the curriculum. He believed a military education should be more than just a technical one of learning mathematics, science, and engineering to apply to practical skills such as fortifications or artillery. He appointed an Episcopal minister, Adam Empie, as West Point chaplain and acting professor of geography, history, and ethics. A swordmaster, Pierre Thomas, was hired in 1814. Thomas conducted the first physical training class for cadets in fencing.

Unfortunately for Swift, the demands of war prevented his plans from taking hold at West Point. The War

Alden Partridge, the Academy's third superintendent, designed the cadet gray uniform. (West Point Museum, Photo by Eilene Harkless Moore)

FREDERICK
TRENCH
CHAPMAN

Department ignored the new law and graduated cadets as quickly as possible with no examinations. Admission standards also were disregarded: One cadet appointed in 1814 was only twelve years old, and both a one-armed cadet and a one-eyed cadet were admitted. Another cadet was married and kept his wife off post, where he would see her at night.

In 1814, an acting superintendent was named, as Swift's wartime duties kept him absent from West Point. Acting Superintendent Captain Alden Partridge graduated from the Academy in 1806 after two years at Dartmouth College. He stayed at West Point first as an assistant mathematics instructor and then as a professor. He left one legacy that is evident to anyone who watches a parade: the iconic cadet full dress uniform.

OPPOSITE: Alden Partridge (left) inspects cadets wearing summer parade uniform in 1815. (West Point Museum, Photo by Eilene Harkless Moore)

CHAPTER 3

Cadet Gray

1814–97

Alden Partridge served at West Point continuously from graduation in 1806 until his dismissal in 1817. He was intelligent and energetic, but also proud, vain, and egotistical. His rigid character prevented him from making any lasting contributions to the Academy, except for one. In the late summer of 1814, he designed a uniform for himself and also one for the cadets—in gray instead of the authorized blue. This gray uniform with modifications is still worn today as the full dress uniform. It forms the iconic image of a West Point cadet.

Partridge wasn't simply being contrary or creative. There was a shortage of blue wool for the Army because the supply of indigo used to dye the wool blue was limited by the ongoing war with Britain. At the time, gray clothing was associated with common laborers, not with "gentlemen cadets," a term widely used in the early 1800s to describe cadets.

Popular legend states that the gray uniform was adopted to honor the valor of the American soldier at the Battle of Chippewa, fought on July 5, 1814, in Canada. But the mundane fact is that gray wool was cheaper than blue, so the War Department authorized the gray uniform in September 1816, largely based on the cost benefit.

General Winfield Scott, the commander of the victorious, gray-clad soldiers at Chippewa against the red-coated British, although not a graduate of West Point, was a strong supporter of the Academy. He retired as commander of the Army in November 1861 and lived at

OPPOSITE: Cadets (left background) and officers (foreground) at West Point, 1832–35. (West Point Museum, Photo by Eilene Harkless Moore)

West Point until his death in 1866. He wrote in his memoirs, published in 1864, that the cadet gray was worn to commemorate the Chippewa victory, and this contributed to the heroic myth for gray, regardless of the accountant's prosaic reality.

The gray uniform authorized in 1816 consisted of a single-breasted coatee made of gray satinette material. Three rows of eight, yellow-gilt bullet buttons in front connected with black silk cord in a herringbone pattern. A standing collar rose as high as the tip of the ear with a button on each side. Four-inch-wide cuffs indented with three buttons along with sewn-on black braid lengthwise were on each sleeve. Three buttons adorned each of the two "swallowtail" style back coat tails.

Cadets used a vest of gray cloth, single breasted, with yellow-gilt bullet buttons, trimmed with black silk lace for winter wear. For summer use a white vest, single-breasted with buttons but without trimmings was worn.

Pantaloons for winter were gray with black silk lace stripes down the sides and a black Austrian knot on the front of each side. Russia sheeting (a type of white cotton) or white jean for summer was worn with the gray coat. A variation of the pantaloon called the "sherryvallies" was worn in 1815, which buttoned up the side pant leg. The pants were worn over the Jefferson shoe. This footwear was like a shoe boot as the top extended above the ankle joint.

Initially the cadets wore tall leather hats called "shakos" similar to those worn by Scott's soldiers at Chippewa. Like many uniform items, the shako was of French origin. French fashion, because of the renown of Napoleon, dominated both the civilian and military clothing of the early 1800s. The first brass cap plate, also modeled on the soldier's cap plate, was worn from 1815–16 and is the earliest distinctive insignia of the Academy.

BOTTOM LEFT:
Stan Gorzelnik (Class of 2011) wears the current full dress uniform with three chevrons, red sash, and sword designating a cadet lieutenant.

BOTTOM RIGHT:
Bullet buttons from the sleeve of a present-day full dress coat reflect the original design of 1816.

Partridge ordered new cadet gray uniforms from a New York tailor, John B. Thorp, made of gray satinette. They arrived in summer 1815.

The common round hat was brought back for use in 1816 because of cost. The large, shiny cap plate was replaced by a plain black cockade with a yellow eagle as the distinctive head gear for a West Point Cadet. A cut and thrust-type sword, yellow mounted with a black gripe, also was adopted.

Partridge always appeared in public dressed in his full dress uniform. He was nicknamed "Old Pewter" by the cadets, probably because he typically wore his personally designed gray uniform instead of the authorized Army blue one. He schemed for more power and went behind Joseph Swift's back to advance a new arrangement for the superintendent with Secretary of War James Monroe in January 1815. Partridge convinced Monroe to approve new regulations that put Swift out of the loop for the Academy's administration. Partridge would report directly to the secretary of war, bypassing the chief engineer. Only the superintendent would commission cadets at his discretion and assign them to a branch within the Army.

Swift reacted quickly to this power grab by Partridge. Swift pointed out to Monroe that Partridge's regulations violated the 1812 law that required the academic staff to set standards for graduation. Monroe accepted revised regulations by Swift in February 1815 that placed the superintendent under the chief engineer, newly designated by Monroe as the inspector of the academy. The superintendent reported only to the inspector and was forbidden to have any direct dealings with the secretary of war.

Swift's lasting contribution to West Point was made with the start of construction of the first stone buildings at West Point in spring 1814. He sited the new barracks, academic building, and mess hall along an east–west line starting on the southeastern boundary of the Plain. The location of the major buildings of the Academy continues to this day

Cadets from 1816–17 in winter parade (left), summer dress (center), and winter undress with gray vest (right). (H. Charles McBarron, Company of Military Historians, Photo by Eilene Harkless Moore)

along the southern area of the Plain, extending to the west. Swift ordered the use of gray granite to construct the buildings, another lasting tradition.

The first barracks constructed for cadets was finished in spring 1815 and was made of locally quarried gray granite. The three-story structure contained long, narrow hallways with forty-eight small, two-man rooms on each floor. Wings at each end consisted of suites for the tactical officers and bachelor officers. A second barracks was finished in 1817 and was located north of the first one. This barracks became known as North Barracks; the other as South Barracks. Together these barracks would house the Corps of Cadets for more than thirty years.

The barracks had no running water and were heated by fireplaces. Water was carried in buckets by plebes, who also carried in firewood from a wood yard. No furniture was provided by the government, so cadets bought their own out of their pay. Although some cadets possessed a bed, table, and chairs, most slept on a wooden pallet, and crates or boxes typically were used for tables and chairs.

Next to the South Barracks was the new academic building, the second structure to be called "the Academy." This two-story building contained the engineering room, chapel, and chemistry laboratory on the first floor. The second floor comprised the philosophical (physics) department, adjutant's office, and the library.

The new mess hall also was a two-story structure with a kitchen and quarters for the mess steward. Cadet mess rooms were on both floors. It was enlarged in 1823 with a back addition for a bigger kitchen and quarters for the staff.

In 1815, Swift retained Partridge as superintendent. But Partridge's overzealous desire for control proved to be his undoing. He was a micromanager who interfered with the way faculty taught their classes. He refused to follow the curriculum of the academic staff. A crescendo of complaints about Partridge from the academic staff and cadets to officials in Washington resulted in a board of inquiry held at West Point from March 16–April 12, 1816.

One of the charges against Partridge was that he and his relatives were defrauding the government. Partridge had appointed his uncle,

TOP LEFT: A cadet shown in 1816 wearing a uniform with the cheaper round hat. (West Point Museum, Photo by Eilene Harkless Moore)

TOP RIGHT: The first cap plate for the leather shako, in 1815, featured a small "y" at the end of "Academy" because the engraver ran out of space for the letters. (West Point Museum, Photo by Eilene Harkless Moore)

Isaac Partridge, mess steward in 1813, when the mess hall reopened after closing in 1807. Isaac Partridge kept a very plain mess called the "commons." George Ramsay, the twelve-year-old admitted as a cadet in May 1814, wrote in his memoirs that "the mess hall was always indifferent, if not positively bad," and he bemoaned its unattractive décor. There were no tablecloths, glassware, or chairs. The tables and benches were painted in red ochre, which stained the cadets' gray coat sleeves and trousers.

Alden Partridge's nephew, Lieutenant John Wright, was commander of the engineer company. Many of the soldiers were neighbors of Partridge from his native Vermont. Partridge allowed the soldiers to cut wood on government land and sell it for a profit. This company, formally called "the company of bombardiers, sappers, and miners," consisted of ninety-four enlisted men. It was established as part of the 1812 act and included the previous detachment of nineteen men formed in 1803.

The board of inquiry considered the charges against Partridge and found him guilty only of improper authorization for woodcutting, though not of fraud. He also was found to be arbitrary and authoritarian toward the faculty and guilty of favoritism among the cadets. He remained as superintendent, but did not change his ways. Because of his interference in the conduct of classes by the faculty and the disruption caused by the court proceedings, no cadets graduated in 1816.

Partridge was a strict disciplinarian who imposed rigorous regulations on the cadets. At 6:00 a.m., they would rise, attend role call formation, and return to clean their room, stand inspection, and then eat breakfast at 7:00 a.m. Mathematics classes were held from 8:00 a.m. to 11:00 a.m. French classes were taught from 11:00 a.m. to 1:00 p.m.

North Barracks, South Barracks, Academic Building (the Academy), and Mess Hall (from left) border the Plain in this 1826 painting. (West Point Museum, Photo by Eilene Harkless Moore)

29

Cadets formed by sections to march to class in 1870s. (Special Collections, West Point Library)

Dinner was from 1:00 p.m. to 2:00 p.m., followed by drawing and engineering classes held from 2:00 p.m. to 4:00 p.m. Drill was conducted from 4:00 p.m. to 5:00 p.m., followed by an evening parade and then supper. Mandatory study in rooms was held in the evenings.

Partridge personally conducted an artillery salute on New Year's Day, 1817. The 1st Company of cadets performed as the firing detail with four cannons. One gun fired prematurely and killed Cadet Vincent Lowe. In 1818, the Corps of Cadets donated $15 per man—a tidy sum at the time—to erect a monument in honor of Lowe. Now located in the West Point cemetery, the monument contains the names of sixty-four cadets and professors who died at West Point up until 1873.

The earliest image of a cadet, from summer 1820, shows an officer with chevrons, sash, and tall plume. In the background are cadets drilling with no colors on the Plain with the Wood monument in center, behind the marching cadets. (Special Collections, West Point Library)

The separation of the duties of superintendent and chief engineer set the stage for the future growth and establishment of the Academy as a premier organization. It would take the right man of character to accomplish this transformation of West Point from a simple mathematics school to the finest engineering school in the world.

President James Monroe visited West Point from June 15–17, 1817. This first visit to West Point by a sitting president was not the cause for celebration that many future presidential visits would be. Monroe investigated the renewed allegations against Partridge by talking directly to the faculty. The President found Partridge to have continued to disregard the academic staff and curriculum. He ordered Partridge to stand trial by court martial and replaced him with Sylvanus Thayer as superintendent.

The thirty-two year old Thayer was well prepared to assume the duties of superintendent on July 28, 1817. He graduated from Dartmouth College in 1807 and immediately attended West Point.

A young Sylvanus Thayer (right) reads a dedication of Wood Monument in 1818 with his staff behind him. (West Point Museum, Photo by Eilene Harkless Moore)

Superintendent Robert E. Lee

Robert E. Lee served as superintendent of West Point from September 1, 1852, to March 31, 1855. As a cadet, he graduated second in the Class of 1829 and is one of only a few cadets to never have received a demerit during his four years at West Point.

Born in Virginia, Lee was the son of "Light Horse" Harry Lee, a Revolutionary War officer and hero. The younger Lee performed brilliantly during the War with Mexico, and General Winfield Scott stated that Lee was "the very best soldier I ever saw in the field."

As superintendent of West Point, Lee earned the respect of all he encountered. General John M. Schofield, who was a cadet under Lee, and subsequently, superintendent (from 1876–81), wrote of Lee that "He was the personification of dignity, justice and kindness, and was respected and admired as the ideal of a commanding officer."

Robert E. Lee, Class of 1829, became a renowned soldier and general, superintendent of West Point, and, later, president of what would become Washington and Lee University. (West Point Museum, Photo by Eilene Harkless Moore)

Lee's experience as superintendent would serve him well when he assumed the presidency of a struggling Washington College located in Lexington, Virginia, after the Civil War. The school grew under his leadership, and is now Washington and Lee University.

Cadets on parade in 1820s with two colors centered in front of lines of cadets and the commandant standing in front of colors. Cannons (right background) fire a salute in front of the post flag pole (right background). (West Point Museum, Photo by Eilene Harkless Moore)

After easily graduating from the Academy in 1808, he served in the Corps of Engineers for two years and then taught mathematics at West Point from 1810–11. His distinguished service during the War of 1812 led to Swift sending him, along with Lieutenant Colonel William McRee (Class of 1805), to Europe for almost two years from June 1815 to May 1817, to study French military schools and fortifications.

Thayer's tour of French military schools had a direct impact on the Academy. Thayer arrived in France in July 1815 just after Napoleon's defeat at Waterloo. The French military schools were closed, so he and McRee used their $5,000 line of credit to buy more than a thousand books, charts, and maps. These items formed the first substantial military library in the United States.

Among Thayer's visits in France was the École Polytechnique in Paris. Renowned for the study of military and civil engineering, the school reopened in early 1816 after recovering from the fall of Paris in 1814. Thayer visited this school, though he did not visit Saint Cyr, the French Military Academy that offered military tactics, strategy, and history in lieu of technical subjects. Thayer also toured the French artillery school of application at Metz. He considered the French approach to higher education, but not the German way. German universities offered electives and emphasized history and philosophy.

LEFT: A button from the first designed cadet overcoat. (West Point Museum, Photo by Eilene Harkless Moore)

FAR RIGHT: A button from the current cadet overcoat with the Academy crest. (Author's Collection, Photo by Eilene Harkless Moore)

A cadet officer wearing a bell-crowned shako, red sash, and sword. (West Point Museum, Photo by Eilene Harkless Moore)

The French method was directive, with all courses prescribed for the students and dominated by mathematics and engineering subjects. Thayer's French-centric thinking would reverberate at West Point into the 1960s. Electives were not offered to cadets until 1960, and the Department of History wasn't formed until 1969.

Thayer collaborated with the faculty to implement a four-year curriculum starting in 1817 for the Class of 1821. The program, drawn from the École Polytechnique, focused on mathematics, science, and engineering. Cadets in their Fourth Class (freshmen) year received instruction in just mathematics and French, with four hours of mathematics and three hours of French. They continued with the same subjects as Third Class (sophomores) cadets. Second Class cadets (juniors) were taught four hours of natural and experimental philosophy (akin to present-day mechanics and physics), two hours of drawing, and one hour of chemistry. First Class cadets (seniors) took four hours of engineering and the military art (history), one hour of chemistry, geology, and mineralogy, and two hours split among geography, history, ethics, and natural law. This curriculum survived almost unchanged until the end of the nineteenth century. This program would initially transform the Academy into the first engineering school in America. But after the Civil War, the now outdated curriculum made West Point a backwater academic college.

Thayer also adopted the French practice of establishing an order of merit based on an objective assessment of cadets in contrast to the favoritism of Partridge. A cadet's ranking in the order of merit determined his future place in the Army. Cadets were graded on a daily basis and the order of merit list published. The first class to graduate based on merit was in 1818, and graduation programs listed cadets by their order of merit until 1978. Cadets were motivated to perform well because they were commissioned into a branch of the Army based on their order of merit, with the highest-ranking cadets entering the Corps of Engineers. Lower ranked cadets went into the artillery, infantry, or cavalry. Cadets still choose branches based on their order of merit.

Another technique Thayer imported from France was the use of small classes called "sections" composed of ten to fifteen cadets. Cadets were organized into sections based on their academic standing. Every Sunday, Thayer received from instructors an academic report on cadets within their sections. The first, or smartest, section advanced faster than the lower sections and covered more subject matter. The last section studied the minimum required to pass the course. Small class sizes and advanced academic courses of required subjects continue to this day.

Fourth Class cadets were organized alphabetically into sections until the January exams and then reassigned based on scholastic achievement. Cadets changed sections every two to three months.

A shortage of faculty forced Thayer to use cadets as assistant professors in order to maintain small class sizes. An average of five or

The commandant's house was built in 1819 by Thayer and is still in use.

six cadets per year mostly taught the Fourth Class mathematics or French. Cadet instructors were paid an extra $10 a month for their efforts and wore fourteen rows of buttons down the front of their uniform coats instead of the usual eight rows. The most famous cadet math instructor was Robert E. Lee (Class of 1829) of Virginia, future General of the Confederate Army during the Civil War. Cadet instructors were employed until the end of the nineteenth century but were reinstituted during 1941–42 when the demands of World War II caused a shortage of officers available to teach.

Grading was based on a point system with 3.0 as 100 percent and 0.0 as total failure. To pass, cadets had to earn a 2.0—any lower grade was deficient. This grading system was last used by the Class of 1980, after which it was changed to the almost universal 4.0 system used by American colleges and universities.

The first diploma was designed in the early 1820s by Thomas Gimbrede, who became the drawing teacher in 1819. He lithographed the diploma and established the first print shop at West Point. This shop published many texts and translations of French works for cadet use in the classroom. The basic diploma design is still used today.

First occupied by Thayer in 1820, the superintendent's quarters have been modified for present-day use.

The Old Cadet Chapel was built in 1836 and moved to he cemetery in 1910 to avoid demolition for new buildings on the Plain.

Another Thayer reform was reduced leave (vacation) time, one of the most unpopular measures he instituted. He abolished the long winter vacations from December 1 through March 15 and the summer leaves allowed by Partridge. Thayer enforced the rule about having cadets report before June 25 to participate in training at a summer camp. The only leave allowed was after the first two years at West Point, the summer after Third Class.

The first diploma was designed by Thomas Gimbrede in the 1820s. (West Point Museum, Photo by Eilene Harkless Moore)

Character development was another component of Thayer's reforms at West Point. While all colleges during the nineteenth century stressed building character as part of their mission, only at West Point would a defective character lead to career disaster as an officer serving in the Army. Thayer attempted to mold each cadet's character by example and by regulations. Thayer's own character was beyond reproach, as evident by numerous graduates' memoirs, some verging on hero worship. To this day, all officers at West Point, whether instructors or staff, are expected to be positive role models for cadets.

In addition to demonstrating character, Thayer established numerous regulations, which were enforced by a newly created Commandant of Cadets. By 1824, cadets were expected to obey two hundred regulations. Regulation number 186 stated that "No cadet shall read or keep in his

The castle-like walls of Thayer Hall, as viewed from Hudson River, exemplify Tudor-Gothic building style started by Superintendent Delafield in the 1840s.

room, without permission, any novel, romance or play." Other regulations forbid throwing snowballs or playing chess.

Unlike Partridge, who personally drilled the cadets and interfered in academic classes by teaching himself, Thayer neither taught nor drilled the cadets. Instead, he delegated daily contact with cadets to the professors and the tactical officers. In September 1817, Thayer appointed Second Lieutenant George Gardiner (Class of 1814), instructor of Artillery, as commandant of the Cadet Battalion. In April 1818, Captain John Bliss, an infantry officer with a distinguished record during the War of 1812, was assigned as the instructor of infantry and Commandant of Cadets. Thayer's regulations of 1825 formally designated the position of Commandant of Cadets as the legal commander of the Corps of Cadets. Congress recognized the position by law in 1858.

Major William Worth served as commandant from 1820–28, the longest tenure to this day. In conjunction with Worth, Thayer

This lithograph from 1857 shows Central Barracks, which were built in Tudor-Gothic style. (Author's Collection, Photo by Eilene Harkless Moore)

established a disciplinary system with precedents that still survive. In early 1825, Worth recommended a demerit system whereby cadets would be "awarded" demerits based on infractions of the regulations. Drinking or being caught out of their room during study period earned more demerits than needing a haircut, which was required twice a month. Demerits were cumulative and were a factor in determining the order of merit for graduation. If a cadet received more than two hundred demerits during the year, he would be expelled. Infractions such as drinking also resulted in additional punishment such as walking extra guard duty, which evolved into "walking the area." Area tours remain the ultimate punishment for cadets. For this punishment, cadets march back and forth between barracks for hours while carrying their rifles, the duration of which is based on the severity of the infraction.

The first company tactical (TAC) officers were assigned by Thayer in December 1820. Lieutenants Zebina Kinsley (Class of 1819) and Henry Griswold (Class of 1815) served as assistant tactical officers, one for each company, and lived in the barracks. Kinsley lived in South Barracks with the 2nd company, and Griswold stayed in North Barracks with the 1st company. To this day, a TAC is the legal commander of each cadet company, and, although they do not live in the barracks, their offices are in their respective cadet company area. Most TACs are West Point graduates.

Initially, cadets selected their roommates, often fellows from their own states. Eventually, to facilitate studying, cadets were required to room by class. Cadets lived in a company area and ate together in the mess hall. A "band of brothers" bond eventually formed among cadets within a class as they roomed together. A need for a symbol of fraternal feelings would lead to the first college class rings designed and worn by the Class of 1835.

Thayer realized the need to present cadets to the public, as most Americans were unaware of the isolated school, still accessible only by boat as late as the 1820s. Worth conducted a series of trips to Philadelphia in 1820, Boston and New York City in 1821, and upstate New York in 1822. These tours were successful in promoting West Point among the citizens of the three largest American cities. The City of Boston presented the Corps of Cadets with their first

This image, from 1842, shows a cadet (left) wearing a shako with artillery insignia, while another cadet (right) wears forage cap. Cadet officers in the background wear red sashes, engineer insignia, and plumes. (West Point Museum, Photo by Eilene Harkless Moore)

colors in 1821. Former President John Adams addressed the cadets when they visited his home in Quincy, Massachusetts. Thayer, satisfied with the favorable publicity, ended the trips in 1822, and cadets resumed summer training only at West Point.

The grand cadet uniform was one of the reasons for the popularity of West Point. In 1818, a bell-crowned shako was adopted, following European fashion. This black leather hat was seven inches high with a semicircular visor and a diamond-shaped yellow plate on the front. A black, eight-inch plume adorned the top, fastened to the hat by a leather cockade, two and a half inches in diameter, with a small yellow eagle. This shako was called "bell crowned" because the top was wider than the bottom so when turned upside down it looked like a bell.

Another Thayer innovation was the adoption of the long overcoat in November 1828. Prior to this, cadets wore civilian coats because a uniform winter coat did not exist. The single-breasted coat was made in a "surtout" (overall) style of gray woolen cloth and extended to within four inches of the ankle. A long cape lined with satinette was attached along the collar. A similar style is worn to this day, although it is double-breasted and extends to just blow the knee. The first distinctive cadet button was designed for this coat with the word "cadet" stamped on the button.

Cadet pay was another area ripe for reform. Before Thayer's tenure, cadets would sell their pay vouchers to civilians at a steep discount to

The present-day dean's quarters was built in the late 1850s in a Victorian style unlike the earlier Federal style of other quarters.

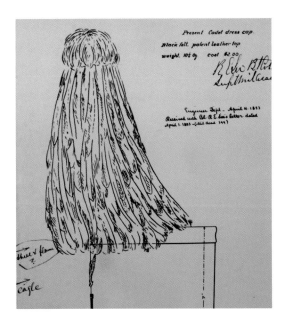

obtain ready cash for contraband items such as liquor. Thayer instituted an internal Academy pay system, with cadets only allocated a certain amount of their pay each month via checks valid only at West Point. This forced savings system ensured that cadets had enough money to buy their Army officer uniforms upon graduation. Variations of this system were used into the 1970s. Cadets were not allowed to bring money with them or have money sent to them while at West Point.

Thayer also initiated construction of new buildings at West Point. He first built quarters for the professors. The oldest surviving building at West Point is the Commandant's Quarters, first occupied in 1819. The superintendent's house is the second oldest structure at West Point, first lived in by Thayer in 1820. Long Barracks burned to the ground in December 1827, the last major building from the late 1700s. Thayer replaced it with a much-needed hotel to accommodate the Academy's numerous visitors. This hotel stood until 1932, when it was demolished.

Thayer enjoyed the complete support of President Madison and his successor, John Quincy Adams. However, Andrew Jackson, elected in 1828 as the first populist President, viewed the Academy as "elitist." Thayer resigned and left West Point in July 1833 after the reelection of Jackson. Thayer was frustrated by Jackson's constant interference in reinstating cadets who had been dismissed.

Superintendents after Thayer carried on his system of both regulations and academics. Those who did want to change the

TOP LEFT: This sketch, drawn by Robert E. Lee, shows a cadet shako, a version of which is still worn today. (West Point Museum, Photo by Eilene Harkless Moore)

TOP RIGHT: Cadets from 1853–61, including a cadet instructor (left), as noted by fourteen rows of buttons and braid instead of the usual eight rows. The cadet in forage cap wears a riding coat while cadet in chair wears full dress. (Frederick T. Chapman and Frederick P. Todd Company of Military Historians, Photo by Eilene Harkless Moore)

curriculum were frustrated by the academic board, which dominated the governance of the Academy until the late twentieth century. The superintendent only had one vote on the board regarding curriculum changes. Success at creating the first and best engineering school in America bred complacency. The strong academic board, which started out as an innovative instrument, became ossified and stilted any change in academic study until well into the twentieth century.

Change did occur in the physical appearance of the Academy under the direction of succeeding superintendents. Rene DeRussy (Class of 1812) served as superintendent from 1833–38. He built the first chapel in the Greek Revival architectural style in 1836. In 1910, the chapel was moved to the West Point cemetery, where it still stands as the oldest public building at West Point. A fire in 1838 burned down the 1815 Academy building and destroyed the early records and papers of West Point. The new academic building was in an Italian Renaissance style.

Richard Delafield (Class of 1818) became superintendent in 1838 and made one lasting contribution to West Point. He built a library in a new architectural style called "Tudor-Gothic" for the Federal Government. This building style, with its massive walls like castles built with gray granite, survives to this day and forms, along with the cadet gray uniform, the iconic image of West Point. The library housed books until 1964. Delafield started construction of a new barracks, also in this style, in 1845 just before he departed West Point for a new assignment.

Robert E. Lee (Class of 1829) was the most famous nineteenth-century superintendent, serving from 1852–55. His lasting contribution to West Point was the design of a new shako as the cadet full dress hat. His sketch of a shako is very similar to the present-day version.

West Point graduates, including Lee, first proved their value in the Mexican War, fought from 1846–48. General Winfield Scott commanded U.S. troops, who captured Mexico City in September 1847. Scott later issued his fixed opinion on the value of West Point stating "that

A cadet captain (left) confers with General Winfield Scott (seated) and officers, ca. 1858–61. (West Point Museum, Photo by Eilene Harkless Moore)

A full dress uniform from 1888 of Officers, Cavalry, and Artillery (from left foreground), a cadet captain (center), and a color sergeant (right). (West Point Museum, Photo by Eilene Harkless Moore)

but for our graduated cadets, the war … probably would have lasted some four or five years …"

Even greater renown was bestowed on graduates after the Civil War. Of the sixty major battles of the Civil War from 1861–65, West Point graduates served as commanders on both sides in fifty-five battles and commanded one side in the remaining five. General Ulysses S. Grant (Class of 1843), the victorious commanding general of the Union Army, was elected President in 1868.

However, success bred even more complacency at West Point, and the Academy drifted behind as knowledge and education dramatically changed other American colleges. The Academic Board saw no need to tinker with the Thayer way in light of the great performance of graduates in both the Union and Confederate armies. The approaching new century and war with Spain in 1898 prompted changes at West Point.

CHAPTER 4

Present-Day Cadet Gray

1898–2000

The short Spanish–American War from 1898–99 prompted reform throughout the American Army as the United States became a world power with the acquisition of the Philippines from Spain. A new superintendent, Albert Mills (Class of 1879), was appointed at West Point. Mills served from August 1898 until August 1906, and many changes occurred at West Point during his tenure, including the creation of a coat of arms, revised full dress uniform, and new construction. The Academic Board still thwarted any attempt at major revisions to the curriculum.

An Academy coat of arms was adopted in 1898. Part of the design included the new motto "Duty, Honor, Country." The helmet of Athena, Greek goddess of wisdom and war, and the American eagle also were featured. Unfortunately, the original design depicted the eagle and helmet facing to the wearer's left. In a tradition older than West Point, Heraldry, from the knights of the Middle Ages, considers facing toward the left as "sinister." After twenty-five years, Academy officials yielded to Heraldry and changed the design in 1923 with the eagle and helmet facing to the wearer's right.

In 1899, the Academy looked to the past for the new full dress uniform, which is worn to this day. The shako was changed to a pattern similar to the version Robert E. Lee designed in the 1850s. An eight-inch pompon of worsted wool adorned the top of the cap. The cap plate was the newly designed coat of arms, also called the crest. Cross belts

The full dress uniform of the early 1900s of a (from left foreground) first sergeant, lieutenant, and adjutant. (West Point Museum, Photo by Eilene Harkless Moore)

The original crest (left) with eagle and helmet facing to wearer's left and the current crest with eagle and helmet facing to wearer's right. (West Point Museum, Photo by Eilene Harkless Moore)

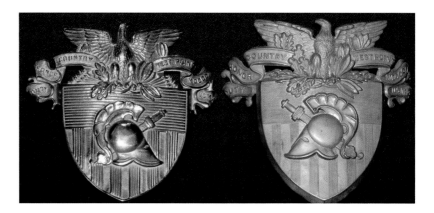

were reintroduced with a black cartridge box placed in the small of the back and held by the white cross belts. A bayonet in a metal scabbard hung from a white waist belt on the left side. A shiny brass plate secured the cross belts and another brass plate served as the belt buckle for the waist belt. The three brass plates provided cadets with plenty of sparkle when marching on the Plain. Cadet officers wore a single shoulder belt across the right shoulder with their sword hanging on the left side. A red sash and tall plume of black cock feathers further distinguished officers.

Toy soldiers on parade wearing the full dress summer uniform were popular for children, as shown by these figures from the early 1920s. (Courtesy of Tom Munson and Craig McClain, Photo by Eilene Harkless Moore)

New uniforms weren't the only changes taking place at West Point. In 1902, Congress approved $5.5 million for major construction, including new barracks, a chapel, academic building, post headquarters, and riding hall. All these structures were designed in the Tudor-Gothic style. The castle-like walls of gray granite, towers, and sally ports contributed to the timeless look of West Point. The chapel resembles a miniature cathedral from England and towers above the Plain. The riding hall was the largest one in America and was used until 1946.

President Theodore Roosevelt attended the Centennial celebration at West Point held from June 9–12, 1902. He arrived via train on June 11 and presented the Medal of Honor to Cadet Calvin Titus in front of the entire Corps of Cadets. Titus performed heroically as a soldier with the 14th Infantry Regiment at Peking, China, on August 14, 1900. He remains the only cadet ever awarded a Medal of Honor. President Roosevelt spoke at the graduation ceremony on June 12 and lauded the contribution of the graduates. Secretary of War Elihu Root spoke words still relevant today when he said, "The Military Academy is more necessary now than one hundred years ago."

Though the Spanish–American War prompted great changes throughout the U.S. Army, World War I resulted in the greatest upheaval at West Point. The War Department graduated the Class of 1917 in April 1917, and then the Class of 1918 just four months later in August 1917. The Class of 1919 was commissioned in June 1918. American casualties from the Army's battles against the Germans starting in September 1918 resulted in the War Department demanding the graduation of the next two West Point classes, 1920 and '21. They both graduated on November 1, 1918; the latter class with only seventeen months of training. This left only the Fourth Class

The riverfront wall of Thayer Hall, originally the riding hall, with the castle-like tower of Taylor Hall looming above it.

ABOVE LEFT: The pinnacle of Tudor-Gothic style is Taylor Hall with its soaring tower. The superintendent's headquarters is located here, along with other administration offices.

ABOVE RIGHT: Cadets walking the area in the snow wearing long overcoats and knit caps for warmth without rifles during the mid-1970s. (Author's Collection, Photo by Eilene Harkless Moore)

(freshmen), admitted in June 1918. A new class was admitted in November to graduate in just a year as West Point was converted to a military training school for officers.

Fortunately for West Point, the war ended on November 11, 1918. Turmoil continued at the Academy, however, as there were no upperclassmen to impart upon the plebes the traditions of West Point and the wartime, one-year course remained in effect. A young, brilliant officer with a fine war record was required. The Chief of Staff of the Army, General Peyton March (Class of 1888), picked Douglas MacArthur (Class of 1903) as superintendent in June 1919.

MacArthur would bring West Point into the twentieth century. The thirty-nine-year-old was one of the youngest superintendents. His outstanding performance during World War I earned him promotion to Brigadier General, a rank he kept when assigned to West Point. His first action was to save the four-year course. Congress had imposed a three-year curriculum after the war to save money. MacArthur, along with numerous other graduates such as General John Pershing (Class of 1886), commander of the American forces in France, advocated for a return to the four-year course. This was accomplished starting with the Class of 1922, most of whom were the cadets admitted in November 1918.

Although the four-year course was saved, the Academic Board prevented major changes to the curriculum. But MacArthur was able to impose changes beyond the purview of the Board. He added one year of graduate school for instructors prior to teaching at West Point. He ordered professors to visit other colleges during the summer to observe civilian teaching methods. He brought in distinguished experts from various fields to present lectures to the Corps of Cadets in order to broaden the outlook of the cloistered cadets living a monk-like existence at their rockbound Hudson home. In addition, upperclass cadets were authorized weekend leave, breaking the Thayer practice of only one vacation during the summer after sophomore year.

Summer training was dramatically changed when MacArthur ordered the cadets to Camp Dix, New Jersey, to undergo training with soldiers in the active Army. Alumni were outraged at the elimination of another sacred cow, as the summer encampment also dated back to Thayer's time.

Unlike Thayer, who had sixteen years to impact West Point, MacArthur only served three years because the new Chief of Staff of the Army, General Pershing, was one of the outraged alumni, and he shipped MacArthur out to the Philippines in 1922. The next superintendent returned summer training to West Point and abolished weekend leave. However, two changes made by MacArthur exist to this day. He expanded athletics among the cadets with an intramural program. Each cadet company is required to field teams in various sports such as football or soccer, and all cadets must participate.

But MacArthur's most important contribution was to create a written cadet honor code and form the Honor Committee, which was run by cadets. The concept of personal honor had always existed at

President Theodore Roosevelt (left in top hat) presents the Medal of Honor to Cadet Calvin Titus in June 1902. (Special Collections, West Point Library)

49

Douglas MacArthur
served as the thirty-
first superintendent
of the Academy, from
1919–22 (West Point
Museum, Photo by
Eilene Harkless
Moore)

UPON THE FIELDS OF FRIENDLY STRIFE,
ARE SOWN THE SEEDS
THAT, UPON OTHER FIELDS, ON OTHER DAYS,
WILL BEAR THE FRUITS OF VICTORY.

West Point, dating back to 1802, when Jonathan Williams sought to impart among cadets the values and honor of gentlemen. A formal code was never defined, nor was an official way to enforce breaches of honor. MacArthur had each company elect an honor representative from the First Class cadets. The chairman of the Honor Committee was the class president. A cadet accused of an honor code violation would appear before an honor board composed of members of the honor committee. A guilty cadet was asked to resign. The case was referred to an Army trial by court martial if the cadet refused to resign, with expulsion as punishment.

MacArthur's saying about the importance of sports for a cadet.

A pamphlet called *Traditions and Customs of the Corps* was written to provide published guidance on the honor code for the cadets. The essence of the Honor Code is distilled in a one-sentence statement still used today: "A cadet will not lie, cheat, steal or tolerate those who do." Cadets enforce the Honor Code to this day.

Two major cheating scandals have marred the Honor Code. In 1951, ninety cadets were dismissed for participating in a cheating ring centered on the football team. A board of officers concluded that there was too much emphasis on football, but the Academic Board stymied changes. A larger cheating scandal in 1976 would finally sweep away the power of the Academic Board and allow the superintendent to impose his will on the Academy.

In spring 1976, another large-scale honor scandal erupted, centered on the Class of 1977, then juniors. Widespread cheating was discovered on a graded homework exercise for the mandatory course in electrical engineering. Eventually, 152 cadets were forced to leave due to the scandal. The unprecedented number of expulsions resulted in a commission formed by Secretary of the Army Martin Hoffmann and headed by former astronaut Frank Borman (Class of 1950). The Borman Commission was highly critical of West Point's administration.

Boxing is required for all male cadets to this day. (Author's Collection, Photo by Eilene Harkless Moore)

Further review of West Point was ordered by Army Chief of Staff General Bernard Rogers (Class of June 1943) by a committee called the West Point Study Group. Their final report, issued in July 1977, ended the dominance of the Academic Board over the superintendent. A revised regulation required the Academic Board to now report to the superintendent not through him to the secretary of the Army.

The result of this new balance of power was unprecedented change at West Point. Academic majors were finally allowed in 1983. This was

The Honor Code on display at Honor Plaza by the entrance to the barracks.

THE CADET HONOR CODE

A CADET WILL NOT LIE, CHEAT, STEAL, OR TOLERATE THOSE WHO DO.

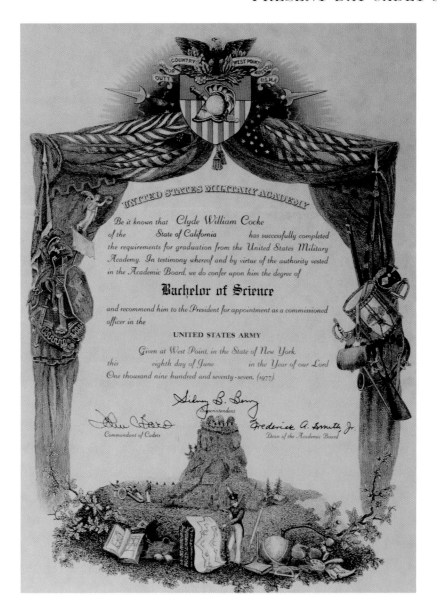

The present-day diploma awards the Bachelor of Science degree to graduates. (Author's Collection, Photo by Eilene Harkless Moore)

the most significant academic change since the Bachelor of Science degree was first awarded in 1933 by an act of Congress.

One reason for granting majors was to secure approval by the Accreditation Board of Engineering and Technology of West Point's engineering programs. Hundreds of electives were offered, and sixteen different majors made available to cadets. This academic program put West Point on equal footing with other colleges that had offered majors for years. The superintendent now had the flexibility to make changes to meet the challenges of the twenty-first century.

CHAPTER 5

Training in Gray and Other Colors

Cadet training at West Point originally consisted of drilling for parades on the Plain. Like much before Thayer, there was no structure or well-organized training before he established the position of Commandant of Cadets in 1817. Williams, the first superintendent, was a scientist, not a soldier. The initial struggle just to survive as a school with no organized curriculum or schedule did not allow for effective training. The law of 1812 establishing the Corps of Cadets with two hundred fifty positions also mandated a three-month summer encampment. Alden Partridge personally conducted infantry drilling and artillery firing training.

Thayer used this law to organize summer encampment, a tradition that would last until 1942. The summer encampment evolved into an elaborate ritual of drill in the morning, cadet socializing in the afternoons, parades, and dances with the ladies of the local cities and towns in the evenings. Uniforms also evolved to allow for the proper dress for various activities. An all-white uniform was adopted after the Civil War for parades.

The first summer encampment during 1818 saw tents erected on the west side of the Plain. The cadets drilled as a battalion with four companies. Expanding the companies from two during the academic year to four in the summer allowed for more leadership opportunities among the cadets. They performed as officers, sergeants, and corporals within the companies. A line of tents was assigned to each

Young Robert E. Lee as a junior officer. (West Point Museum, Photo by Eilene Harkless Moore)

Cadets conduct artillery drill on the Plain in the 1820s, with the Wood monument in the foreground. (West Point Museum, Photo by Eilene Harkless Moore)

company. Each tent had a wooden floor but no cots as the cadets slept on the floor.

The Commandant and other officers lived on one end of the camp. Cadets performed guard duty twenty-four hours a day. They also served as Officer of the Day and would dine with Thayer as part of this duty. Cadets started their day with an inspection in ranks under arms

Colonel Thomas Neill, commandant, at summer camp seated (second from right) with tactical officers in 1879. (Special Collections, West Point Library)

(carrying rifles and bayonets). A daily parade was held in the late afternoon, followed by supper.

Artillery drill was conducted with cannons. Cadets moved the cannons by pure muscle until 1839, when horses were finally authorized for West Point. Between drill and parades, cadets kept busy cleaning their muskets, swords, bayonets, white belts, and uniforms.

The summer encampment on the Plain had evolved by June 1825, when a new cadet named Robert E. Lee reported into a well-organized operation. The camp was now conducted on the northeast part of the Plain where Fort Clinton once stood. This location was away from the barracks and quarters.

By this time, the Academic Board, as part of Thayer's reforms, first conducted an oral entrance examination before candidates were allowed to begin training. Cadets would not sign the oath of office until after completing summer training, which signified acceptance into the Academy. Their cadet warrant (formal appointment) was not given until the end of the first exams in January.

Twenty applicants failed the entrance exam that Lee easily passed. Lee brought with him to West Point at his own expense in the required leather trunk two pair of Jefferson shoes (shoe boots), two pair of white leather gloves, seven shirts, seven pair of wool socks, seven pair of cotton socks, four pocket handkerchiefs, six towels, two pillow cases, two

Cadets conduct artillery drill in 1875 in gray fatigue jackets at the guns supervised by an upperclass cadet in full dress (right middle ground). (Special Collections, West Point Library)

blankets, a clothes bag, a hair brush, a comb, and a tooth brush. He was issued the following clothing: one uniform full dress coat, one gray cloth vest, two pair of gray pantaloons for winter, four pair of white pantaloons for summer, two black silk stocks, and two sets of white belts. A stock was worn around the neck under the collar.

Other items for summer camp included a mattress, pillow, mirror, chair, candlestick, tin wash basin, water pitcher, tin cup, broom, and, most importantly to the Academy, an account book. Each new cadet bought all these items from the government out of his pay of $16 and subsistence allowance of $12 a month. His expenses were entered into his account book. The Academy treasurer also kept an account for a cadet's monthly pay, which would be reconciled with the cadet's account book each month.

Lee paid $8, half a month's pay, for his full dress cap, the bell-crowned leather shako. The cap was heavy, weighing five pounds, and was uncomfortable to wear. Lee designed his own shako as superintendent during the 1850s, which was lighter than the one he wore as a cadet.

An 1890s cadet wears dress gray while jumping a five-foot hurdle. (Special Collections West Point Library)

Among the earliest uniform variations were fatigues, which were worn to prevent wear and tear on the full dress uniform. Cadet Lee also bought one pair of blue cotton fatigue pantaloons, one blue cotton fatigue jacket with sleeves, and a gray cloth forage cap for wear at summer camp. The forage cap replaced the civilian round hat. Designed to be turned upside down to hold food items such as eggs, hungry soldiers used the cap to take food back to camp.

The arrival of horses in 1839 coincided with the admission of one of the finest horsemen ever to attend West Point. Ulysses Grant (Class of 1843) was an indifferent cadet, but he excelled at riding. The riding master assigned Cadet Grant a strong-willed horse named York. During graduation ceremonies for the Board of Visitors in 1843, Grant jumped the horse over a pole standing more than six feet high, a record that lasted for twenty-five years.

Horse training resulted in new uniforms for riding and summer training. A gray riding jacket was adopted in 1849. Buckskin gauntlets (leather gloves) and a black leather belt were authorized in 1850. A gray shell jacket was designed as the fatigue coat. In 1861, a "chasseur" style blue forage cap was introduced. This hat, also of French origin, was similar to the one worn by Union troops during the Civil War. White linen jackets and trousers were issued in 1870 for summer camp wear. A "white India helmet" was introduced during the summer camp of 1878. This cork helmet was styled after the British Army

Camp depicted in the 1850s in a lithograph. (Author's Collection, Photo by Eilene Harkless Moore)

Lieutenant Ulysses Grant (right) wearing a forage cap holding his horse in 1845. (Special Collections, West Point Library)

ABOVE: Model of a cadet wearing the riding uniform introduced in the 1850s. (West Point Museum, Photo by Eilene Harkless Moore)

ABOVE RIGHT: A well-worn cadet "chasseur" style forage cap of the late 1800s with an embroidered cap insignia. (Special Collections, West Point Library)

Cadets wearing the riding uniform formed up on the Plain in the late 1870s for cavalry drill. (Special Collections, West Point Library)

helmets worn in India to protect the wearer from the sun, not from bullets. The Franco–Prussian War of 1870–71 ended in the defeat of France and established Prussian–German military dominance. Military fashion soon reflected this change, so in 1881 a Prussian-styled spike was added along with a gold chain to the white helmet.

A gray coat trimmed down the front center, around the bottom, and up the back with a black mohair braid one inch wide was adopted in 1889 to replace the gray riding and shell jackets. This full-collared coat is worn today as the semiformal uniform called "dress gray." The turn-down white collar was replaced in the same year with a starched collar, attached inside the coat collar to show three eighths of an inch outside. This collar is still worn, much to the discomfort of cadets.

Neither uniforms nor training advanced to keep pace with modern warfare. Cadets were still using muzzle-loading cannon from the Civil

A cadet wears a white linen uniform with spiked helmet (center) and a cadet lieutenant (right) in full dress at summer camp during the 1880s. (West Point Museum, Photo by Eilene Harkless Moore)

A guard (left background) on duty at the edge of the camp during the late 1870s as other cadets are dressed for leisure time. (Special Collections, West Point Library)

Cadets relax at camp during the late 1870s. One cadet wears the riding uniform (left standing) while the others are in various stages of full dress in front of a Company D tent (sign above center tent pole). (Special Collections, West Point Library)

War era in the 1890s. Captain Otto Hein (Class of 1870) became Commandant in June 1897. He updated cadet training to reflect what he learned from the German approach to training. Hein spent five years as the military attaché in Vienna, then the capital of the Austro-Hungarian Empire, and observed how European armies trained.

He first issued to the cadets a field service uniform in 1898 with Army standard field equipment. This uniform was of cadet gray, and the gray color was worn until World War II. Cadets wore a gray campaign hat, gray woolen shirt with tie, gray trousers, leggings, and full Army field equipment. Items included canvas looped cartridge belt, blanket roll, entrenching tool, and canteen.

This uniform replaced the all-white linen version most suited for the parade ground. Cadets still learned parade drill, but Hein added hands-

Cadets "rally to the colors" during the 1880s in a drill to protect the flag from cavalry, already rendered obsolete by longer-range rifles. (Author's Collection, Photo by Eilene Harkless Moore)

Cadets wearing dress gray conduct gun drill in the early 1900s. (Special Collections, West Point Library)

on training in open-order drill, tactical formations, and reconnaissance. The white summer helmet was discontinued in June 1903.

Another uniform modification was adopted on October 24, 1899. Service stripes, one for each year, were added to the cuffs of the full dress and dress gray coats to more easily identify the status of a cadet.

New rifles also were introduced. The first magazine rifle, the Krag-Jorgensen, was used by cadets in 1895. In 1905, the Springfield Model 1903 rifle was issued to cadets. Cadets trained with this rifle until 1942 when the M-1 Garand semiautomatic rifle was issued. In 1963, the M-14 rifle was introduced. When the M-16 rifle, made with black plastic, was issued to the Army during the mid-1960s for the jungles of Vietnam, West Point retained the wooden stock M-14 because it looked better for parades. Cadets still parade with the M-14 rifle but train with the M-16.

Cadets wearing the field service uniform pause for a drink during the 1920s while the cavalry detachment of Buffalo soldiers ride in the background. (West Point Museum, Photo by Eilene Harkless Moore)

Hein set the stage for subsequent Commandants to improve training. Rifle marksmanship training and a practice march were added to the training regimen. The march allowed cadets to move out from the confines of West Point and live in tents as soldiers did in the Army on maneuvers. First Class cadets assumed officer roles during this extended field training exercise, which lasted up to two weeks.

Not all changes of this era were for the better. One negative practice started during summer camp was the hazing of new cadets, nicknamed "plebes." Before the Civil War, upperclassmen—seniors, or "firsties," and sophomores, or "yearlings"— played practical jokes on plebes while they slept or performed guard duty. A favorite trick was to hide a sleeping new cadet's clothes so he would have to report to morning formation wrapped in a blanket.

Unfortunately for West Point and the cadets, this relatively harmless practice of humorous pranks escalated after the Civil War

Cadets conduct skirmish drill, a more open-order drill, wearing dress gray on the Plain in early 1900s. (Special Collections, West Point Library)

into a systematic program of physical hazing. New cadets began their training by first undergoing a three-week indoctrination in the barracks with a cadre of cadets and officers. This atmosphere allowed the upperclass cadets to haze the new cadets. The harassment continued once the new cadets entered the summer camp. The new cadet training was called "Beast Barracks" by the cadets, a nickname that endures to this day for cadet basic training.

New cadet Douglas MacArthur reported to West Point in the summer of 1899 during the most intense period of hazing. MacArthur received extra attention because his father, Arthur MacArthur, was a Medal of Honor winner from the Civil War and an Army General. While in summer camp as a plebe, MacArthur was forced to perform deep knee bends for more than an hour until he collapsed. MacArthur later testified before a Congressional committee about hazing. The negative consequences of hazing were exemplified by Cadet Oscar Booz, who entered West Point in summer 1898 and was

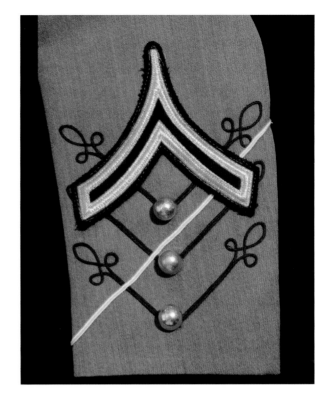

A single stripe on the sleeve indicates a sophomore, while the single chevron on the sleeve denotes a corporal. (Author's Collection, Photo by Eilene Harkless Moore)

Cadets of the early 1900s wearing field service uniform train on the march with field artillery. (Special Collections, West Point Library)

A cadet perception of "Beast Barracks" from the 1951 yearbook. (Author's Collection, Photo by Eilene Harkless Moore)

subject to unrelenting hazing, including drinking bottles of Tabasco sauce at meals. He resigned in October 1898 with his throat damaged from excessive drinking of the hot sauce. He died in December 1900 from tuberculosis, which his father blamed on West Point. The outrage from the news media caused Congress to pass a law in 1901 that outlawed hazing.

Hazing subsided prior to the outbreak of World War I in 1914. The war also saw some new uniforms at West Point. When the new plebes reported in November 1918, they wore the Army olive drab enlisted uniform to distinguish them from the plebes in gray of the

summer of 1918. They also wore an orange hat band, earning them the nicknamed "the Orioles." When the members of the early graduating class of November 1918 were recalled to West Point after the end of combat in November 1918, they wore the Army officer uniform and were considered "officer students" until graduation in June 1919.

The aftermath of World War I resulted in the new superintendent, MacArthur, imposing drastic changes in training. The firsties and yearlings spent the summer of 1920 at Camp Dix, New Jersey, training with the regular Army. They learned the tactics and trained on the weapons of World War I while interacting with soldiers. The new cadets were trained by the tactical officers in a positive way to totally eliminate hazing. However, a clamor of criticism from all sides ended this experiment after only two summers. The old graduates called the

Drills such as these performed by new cadets during the 1880s were abused and led to hazing. (Author's Collection, Photo by Eilene Harkless Moore)

67

Cadets wearing khaki drill uniforms are inspected during the summer of 1955 while the band plays in the background. (West Point Museum, Photo by Eilene Harkless Moore)

Summer encampment on the Plain in the early 1900s. (Special Collections, West Point Library)

Cadets wearing full dress stand inspection in ranks during summer encampment in the early 1900s. (Special Collections, West Point Library)

new cadets soft because they had not endured hazing. The officers' wives complained about the end of the grand social life of the summer camp dances and parades.

It would take the Second World War to finally end the summer encampment on the Plain. New land purchased earlier allowed for modern training. The plebes stayed in the barracks at West Point the entire summer of 1943 while the yearlings trained at the new Camp Popolopen. With more than 10,000 acres of land, there was ample space for shooting modern weapons such as artillery and for conducting combat maneuvers. The camp was renamed Camp Buckner after the war in honor of Simon Buckner, commandant in the 1930s, who was killed in action on Okinawa in June 1945.

New uniforms were adopted to reflect the modern training. A gray coverall uniform replaced the cadet field service one in 1938. In 1942, the Army steel helmet, combat boot, field jacket, and khaki drill uniform were issued for training. In 1946, the Army green fatigue uniform was first used for cadet training, and cadets now use the current Army combat uniform for training.

FREDERICK
TRENCH
CHAPMAN

CHAPTER 6

Hellcats and Buffalo Soldiers

The Soldiers of West Point

The West Point Band, the oldest Army unit at West Point, is composed of soldiers, not cadets. There have been "field music" musicians with drums and fifes since General Samuel Parsons's 1st Connecticut Brigade established West Point as a permanent post in 1778. More than sixty fifers and drummers accompanied this brigade. During the eighteenth century, drummers and fifers were vital to a commander on the field of battle in order to control soldiers in combat. Commands were relayed to soldiers via the drum and fife, a role today performed by radios. Bugles were added during the nineteenth century because of their greater sound range.

A military band in the 1800s consisted of musical instruments such as bassoons, bugles, clarinets, French horns, cymbals, bass drum, flutes, and trumpets. The law of 1812 establishing the Corps of Cadets authorized four musicians per Cadet Company. A civilian teacher of music was hired in 1816 at West Point with the pay of a major. The first teacher of music was Richard Willis from Ireland. An accomplished bugler, he established a tradition of musical excellence that continues to this day. Present-day musicians must audition to be accepted into the band, and 97 percent have a Bachelor's degree in music, with 58 percent also having a Master's degree and some even holding a Doctorate in music

In 1817, the musicians were formally named the West Point Band, which is the oldest U.S. Army band. There were sixteen musicians in

The drum major with his baton, red sash, and sword leads the band in 1818. (West Point Museum, Photo by Eilene Harkless Moore)

Present-day Hellcats wearing their parade uniform, called "full dress over white," worn when the cadets also wear their full dress over white.

Hellcat drummers and buglers wearing the Army combat uniform play during the lunch formation to march the 3rd Regiment into the mess hall while Thayer looks on in the background.

the military band and six field musicians of the field music element. The strength of the band has varied over the years based on funding from Congress. Cadets partly paid for the band by a monthly payroll deduction of twenty-five cents per cadet from 1816–98. Present-day strength is seventy-nine musicians and three officers.

Cadets nicknamed the field music element "Hellcats" during the early 1900s because they awoke every morning to the din of drums and bugles courtesy of these musicians. Today the Hellcats are most visible at the lunch formation when a detachment plays as the cadet 3rd Regiment marches into the mess hall. There presently are thirteen field musicians in the Hellcats. They play rope snare drums similar to those of their Revolutionary War predecessors' drums. The bugles are actually field trumpets but are called bugles in keeping with tradition. (A bugle is limited to one musical key while field trumpets allows for

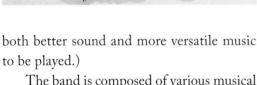

FAR LEFT: Musicians of the band, from 1822–31. (H. Charles McBarron Company of Military Historians, Photo by Eilene Harkless Moore)

LEFT: Musician wearing the Thayer-designed 1822 uniform of white trimmed in red. (West Point Museum, Photo by Eilene Harkless Moore)

both better sound and more versatile music to be played.)

The band is composed of various musical ensembles, including the concert band, Hellcats, marching band, Jazz Knights, and chamber ensembles. The marching band is formed by members of the other sections and plays for all cadet reviews and during half time for home football games in their unique blue uniforms. Sixty-two musicians usually play for a cadet review.

The band first started to wear a special uniform authorized by Thayer in October 1822. Prior to Thayer's tenure, the band wore redcoats, the reverse color of the Army's usual blue with red facing (trim on cuffs and collars). The reverse colors date to the American Revolutionary War when commanders needed to easily identify drummers and fifers in the smoke of battle to relay their orders to soldiers. The Thayer uniform consisted of a coat similar to the cadet's full dress but in white cassimere uniform. The collar, sleeves, and coat tails were trimmed in red. The

The band during the early 1860s wore blue, and the drum major wore a French-style bearskin hat. (Special Collections, West Point Library)

For a short time from 1867–75, the band again wore white trimmed with red. (Special Collections, West Point Library)

white pantaloon also had a red stripe. The bell-crowned shako, army sword, and red and white plume completed this uniform.

In May 1850, the band adopted the Army blue uniform. In 1867, a white uniform also trimmed in red was introduced, but it only lasted until 1875 when blue was again issued. During the 1880s, the band wore a Prussian-style spiked helmet with a black or white horsehair plume. For the West Point Centennial, the band adopted the uniform worn to this day with minor variations. The dark blue high-collar uniform coat is made at the Cadet Uniform Factory located at West Point. The hat is the cadet full dress "tarbucket" with a white cord and distinctive band cap plate. The drum major wears a white plume of cock's feathers and a baldric (sash) with miniature drumsticks.

The band in the late 1870s wearing blue uniforms, with the director on the far left in an officer's uniform. (Special Collections, West Point Library)

The drum major wore a distinctive bearskin hat with other members of the band in 1872. (West Point Museum, Photo by Eilene Harkless Moore)

Along with the band, detachments of soldiers also have been stationed at West Point to assist in training the cadets and performing the support functions necessary for an Army post. The first detachment of just nineteen soldiers from the Corps of Engineers in 1803 was insufficient. In April 1812, the Company of Bombardiers, Sappers, and Miners was raised and incorporated the original detachment. This company of ninety-four men was disbanded by Thayer in 1821 and replaced by an Artillery company. In 1829, Thayer received permission from the War Department to organize a detachment of artillery to support West Point. Various detachments were organized as needed to assist with training the cadets.

In 1839, a cavalry detachment arrived. A sergeant, five dragoons (horse-mounted soldiers), and twelve horses started training the cadets. A new detachment of cavalry arrived in March 1907 composed of eighty-

The band during the late 1870s at summer camp (background) also wore white pants as a summer uniform. (Special Collections, West Point Library)

nine troopers (mounted soldiers) of the 9th Cavalry. This unit was composed of African-American men who had earned an outstanding reputation on the western frontier fighting the Indians. The 10th Cavalry also consisted of African Americans, and both units were given the title "Buffalo Soldiers" by the Indians to honor the warrior spirit of the men whose black hair resembled the buffalo, considered a sacred animal by the Indians. In February 1932, the 9th Cavalry was redesignated the 10th Cavalry Detachment until it was disbanded on September 1, 1947, as horses finally gave way to tanks.

The West Point Band, wearing a summer uniform, plays at summer camp during the late 1920s before cadets wearing dress gray over white. (Special Collections, West Point Library)

Present-day band members in their summer uniform, called "Sierra," worn when the cadets wear white over gray. (West Point Band)

Artillery soldiers shown in full dress with a captain (left) wearing red plume and sash during the late 1830s. (H. Charles McBarron Company of Military Historians, Photo by Eilene Harkless Moore)

Other companies also soon were approved. An engineer company of Sappers, Miners, and Pontoniers was authorized by Congress in May 1846 with one hundred men. Pontoniers constructed temporary bridges using pontoons. These soldiers deployed to Mexico in 1846 and served there until the end of the War with Mexico in 1848 when they returned to West Point. The company stayed at West Point until the Civil War began in 1861, when they deployed with the Army of the Potomac and returned to West Point in 1865. In 1902, the Engineer Company left West Point, and a detachment remained until 1954, when the engineers were included in the Combat Arms Detachment of

Dragoons shown wearing full dress with mounted trumpeter in a red coat during the late 1830s. (H. Charles McBarron Company of Military Historians, Photo by Eilene Harkless Moore)

The cavalry detachment composed of African Americans known as the Buffalo Soldiers during the early 1900s. (Special Collections, West Point Library)

A Buffalo Soldier bugler wearing the olive drab service uniform (right foreground) confers with an artillery officer (center) and the superintendent (left foreground) in 1941. The officers are wearing the blue mess uniform for an evening formal function. A nurse wears service dress with a blue coat (left background), and cadets are (right background) in full dress. (U.S. Army print, Photo by Eilene Harkless Moore)

the Department of Tactics. This detachment was disbanded in the 1960s.

After Word War II, the various detachments, including female nurses, were consolidated as the 1802nd Special Regiment. Part of this regiment included a military police detachment. The MP's mission was

Engineer soldiers wearing winter full dress during the late 1840s with an officer (right) in dark blue. The soldiers wear the same shako and cap plate as the cadets. The sergeant (left) wears a tall plume as a mark of rank. (H. Charles McBarron Company of Military Historians, Photo by Eilene Harkless Moore)

to enforce the laws and regulations at West Point. In 1958, the MP company was attached to the 1st Battalion, 1st Infantry Regiment, which had replaced the 1802nd Special Regiment in 1956. In 2007, the 1st Battalion was deactivated, and the MP Company is now part of the newly formed Army Installation Command, which manages all Army installations. The MPs are now called the U.S. Army Garrison MP Company and also perform ceremonial duties such as forming the firing party for funerals held at West Point.

Women nurses at West Point during the 1950s with the officer (right) wearing the summer dress uniform and the soldier (left) wearing the summer service uniform. The officer wears on her left shoulder the patch of the 1802nd Special Regiment. (Author's Collection, Photo by Eilene Harkless Moore)

Present-day soldiers of the Military Police Company wearing the Army Service Uniform stand ready to perform as a firing party at the West Point Cemetery.

CHAPTER 7

A Uniform for Every Occasion

From the Classroom to the Ballroom

Cadet uniforms have evolved over the years to meet the growing number of different cadet activities. Originally, cadets wore their full dress coats to class and for all meals. Fatigues were worn in the summer and for work details.

That changed in the early 1840s when Superintendent Richard Delafield adopted a furlough uniform for the juniors. A blue frock coat similar to a civilian garment, with a double row of cadet brass buttons, was prescribed, along with a white shirt and white trousers. Cadets also wore an officer's undress kepi cap or a civilian straw hat.

In 1848, West Point issued the Army officer blue frock coat without rank insignia to cadets. The juniors acquired the nickname of "cows"—used to this day—from the way they ran across the Plain when returning from their summer furlough. When the cadets ran across the Plain en masse on the reporting day back from furlough, an observer remarked they looked like cows returning to the barn.

Cadets wore dress gray to class instead of full dress starting in the 1890s. After World War II, a class uniform was designed with a dark blue shirt with black tie. In the 1970s, a short-sleeved classroom shirt with an open collar was adopted for early fall and spring wear. A gray windbreaker jacket was issued in 1946 for wear to class with the class uniform. The gray jacket was replaced by a black jacket in 2006.

A short overcoat was authorized in 1926. This double-breasted coat was without a cape and extended to just below the waist; the Class

Plebes wearing the class uniform march to class (left) while an upperclassman wearing a jacket (right) with the "A" for playing on a varsity sport walks to class during the 1950s. (West Point Museum, Photo by Eilene Harkless Moore)

The short overcoat was worn in the mid-1970s to class during the winter, along with the garrison cap. (Author's Collection, Photo by Eilene Harkless Moore)

of 2011 was the last to be issued the short overcoat. A black wool hooded parka was first worn in 1952. Cadets now wear either the jacket or parka instead of the short overcoat with the class uniform.

A physical training uniform specifically designed for gym use was not authorized until 1953 when gym shorts and a T-shirt, with the Academy crest were approved for use. Previously, cadets performed their numerous physical activities in variations of their uniforms.

The Academy's most famous athletic gear is the football uniform. Dennis Michie, Class of 1892, organized, played in, and coached the first football game at West Point on November 29, 1890. The game was played on the Plain, and Army lost to Navy: 24 to 0. Navy already had been competing in football for eight years. The next year, Army beat Navy at the Naval Academy: 32 to 16. The event has since evolved into one of the greatest rivalries in college football. It is usually played in Philadelphia and broadcast on national television. Both the Corps of Cadets and the Brigade of Midshipmen march onto the field before the game. In March 1899, the colors black, gray, and gold were selected for intercollegiate athletic uniforms.

Michie was killed in action during the Spanish–American War at the Battle of San Juan Hill in 1898. The football stadium, completed in 1924, was named after him. Today, Michie Stadium is rated by national sports magazines as one of the top ten venues in America for watching college football.

Officers in blue frock coats (foreground) along with civilians enjoy strolling on the Plain during the late 1850s. (Author's Collection, Photo by Eilene Harkless Moore)

Plebes taking boxing lessons wearing t-shirts and gray trousers in the late 1940s. Gym uniforms were not provided until 1953. (Author's Collection, Photo by Eilene Harkless Moore)

During the 1899 game, held in Philadelphia, Navy appeared with a goat as their mascot. A quick-thinking Army quartermaster officer procured a mule from the streets of Philadelphia, and a mule has been the mascot ever since, with the first official mascot a pack mule in 1936. Mules were used in the Army to carry supplies and even disassembled artillery guns called mountain artillery. Present-day mules are donated

The first Army–Navy game was played on the Plain on November 29, 1890. Cadets (background) watch while wearing their long overcoats. (Special Collections, West Point Library)

A football player
from 1902 wears the
uniform with a black
knit sweater and letter
"A." (West Point
Museum, Photo by
Eilene Harkless
Moore)

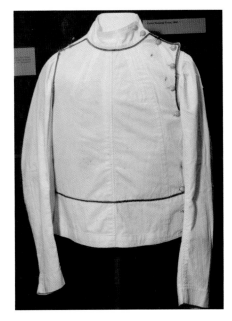

TOP LEFT: Present-day mules Ranger II (right) and Raider (left) greet fans during a football game at Michie Stadium.

LEFT: Piper Richard McPhee (Class of 2013) wears the official West Point tartan with a modified dress gray coat.

BOTTOM LEFT: Fencing uniform coat of the early 1900s. (West Point Museum, Photo by Eilene Harkless Moore)

The 1900 fencing team. (West Point Museum, Photo by Eilene Harkless Moore)

to West Point, and the cadet mule riders volunteer to ride at home football games and, of course, at the Army–Navy game. At that 1899 game, the cadets beat the midshipmen: 17 to 5.

Though football is hugely popular at the Academy, twenty-five intercollegiate sports are played at West Point, with women's teams in nine sports. Club sports comprise another twenty-eight teams but are sponsored by the Directorate of Cadet Activities.

The Directorate of Cadet Activities sponsors 115 clubs, and 70 percent of cadets belong to at least one club. The myriad groups range from Astronomy to Wargames Committee.

Of all of West Point's clubs, the Pipes and Drums of the U.S. Corps of Cadets wear the most colorful uniform. The pipers wear a tartan registered by the Court of the Lord Lyon, King at Arms of Scotland. West Point's colors of black, gray, and gold form the pattern for the tartan. In 1985, Superintendent Willard Scott recorded the tartan with the Lyon court. This is one of the few American tartans registered in Scotland as an official tartan.

The first uniforms designed for an athletic team were worn by West Point's Fencing team. The white jacket dates back to the 1800s. Fencing was the first sport taught by the original Sword Master in

THIS PAGE: The drill team performs on the Plain before a football review.

OPPOSITE TOP LEFT: Cadets attend a hop (dance) during the 1880s. (Author's Collection, Photo by Eilene Harkless Moore)

OPPOSITE TOP RIGHT: A black blazer coat was adopted in 1968 and worn until 1997. (West Point Museum, Photo by Eilene Harkless Moore)

OPPOSITE BOTTOM: Cadets in full dress dancing at a Christmas hop in 1898 at West Point. (Special Collections, West Point Library)

1814. The title was changed to Master of the Sword in 1881, and it is still used by the director of Physical Education. The Fencing team is now a competitive club sport.

One of the most visible club sports teams is the Sport Parachute Team. There are thirty-six members of the team called the Black Knights. Selected cadets jump in the game ball for all home football games at Michie Stadium. They also jump onto the Plain prior to home football game parades.

Another club that performs prior to home football game reviews is the Black Knight Drill Team. Twenty cadets comprise this exhibition rifle drill team, which performs with M1903 Springfield rifles tipped with bayonets. They perform in dress gray coats with white trouser wearing white belts.

In addition to sports, dancing has been an important social event at West Point for the cadets. Dance classes were conducted during summer encampments and also for many years after World War II. Weekly dances called "hops" were held through the 1980s. Dances are

A collection of hop cards from the 1860s (middle foreground) to the 1960s (right foreground). (Special Collections, West Point Library, Photo by Eilene Harkless Moore)

still held for events such as graduation, usually at the cadet activities center, Eisenhower Hall.

Initially cadets wore full dress to dances, but as society relaxed standards of dress, so did West Point. Dress gray became the usual uniform for weekly hops, and full dress was worn for special occasions like graduation. Reflecting the relaxed dress standards of the 1960s, a civilian-type uniform was adopted in 1968. A black blazer with class crest, white shirt, tie, and pants were issued to cadets until 1997. A cadet casual uniform of polo shirt and pants is now authorized. Cadets select from several different colors of shirts.

Hop cards also are a reflection of a bygone era. Cadets would use these dance cards to arrange with young ladies the order in which they would dance with them. Social etiquette called for the cadets to dance with a number of women, not just one. This practice started in the mid-1800s and lasted into the late 1960s.

Another event that calls for a special uniform is the Army–Navy game. Cadets wear the latest Army Combat Uniform to class the week prior to the game in lieu of the class uniform. Cadets also wear the ACU for various club activities such as the Wargames Committee.

ABOVE: Cadets wearing the ACU engage in the Osprey game "Force on Force" during the Pointcon convention in April 2011.

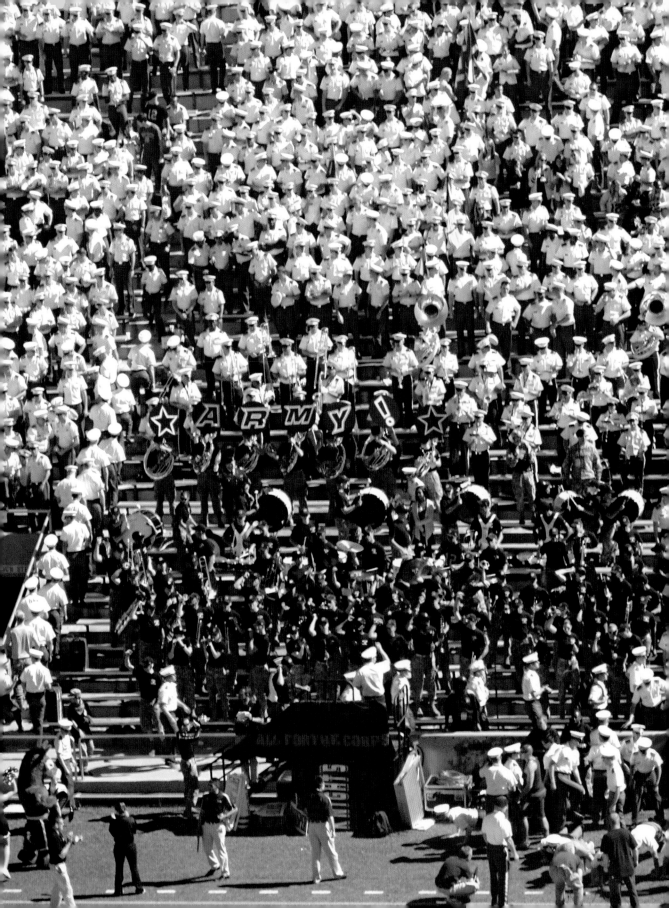

CHAPTER 8

Companies of the Corps

From 2 to 36

Cadets have been organized into companies since the Congressional Act of 1812 mandated their formation. The law did not specify the number of companies, and two were formed in 1812, simply called 1st and 2nd companies. To increase cadet training and leadership opportunities, Thayer ordered four companies formed only for summer training starting in 1818. The four companies were then kept for the entire year starting in April 1827.

In 1815, the U.S. Army adopted an alphabetical system to identify companies within a regiment. West Point did not change to lettered companies until 1831, when the companies were renamed as A, B, C, and D. Four lettered companies were used until 1900 when the growing size of the Corps of Cadets resulted in the formation of E and F companies. The authorized strength of the Corps of Cadets in 1900 was 481 cadets organized within the six companies.

The advent of World War I in 1914 caused Congress to authorize 1,336 cadets in 1916, one year prior to the entrance of the United States into the war in 1917. The Corps of Cadets was then expanded to nine companies on June 8, 1917, to accommodate the increased strength with the formation of G, H, and I companies. Three more companies, K, L, and M, were activated on July 19, 1919. Following Army practice, J was not used as a lettered company because in the early 1800s, the letters I and J could be confused in hand-written orders, so only I was used as a lettered company.

Cadet Company guidons are carried to the football stadium. The cadet spirit band in black stands out among the white over gray uniforms.

Congress expanded the Corps of Cadets in 1935 to 1,964 members, but there still remained twelve companies. World War II again prompted an increase in the number of cadets, to 2,500, in June 1942. A second regiment was formed with the enlarged Corps of Cadets now comprised of two regiments of eight companies each, with A-1 through G-1 in the 1st Regiment and A-2 thru G-2 in the 2nd Regiment.

Starting in academic year 1946, companies H, I, K, L, and M were added for each regiment to form twenty-four companies in two regiments for a total of 2,500 cadets. This organization remained in effect until 1965, when four regiments of six companies each were organized in preparation for expansion of the Corps of Cadets to 4,400.

In 1964, President Lyndon Johnson signed legislation for the largest-ever expansion of the Corps of Cadets. President John F. Kennedy initially approved the proposal for more cadets based on a conversation with Superintendent William Westmoreland (Class of 1936) during the Army–Navy game in December 1962. Kennedy asked Westmoreland why there were there so many more Naval midshipmen at the game than cadets. Westmoreland pointed out that the Navy was authorized two thousand more men, and the President agreed to support an expansion to match the Navy. The number of companies gradually increased to thirty-six in 1969 based on the construction of new barracks and other facilities. Until 1998, each regiment consisted of nine companies (A–I).

Congress reduced the size of the Corp of Cadets in 1993 by 10 percent as part of the reduction in the Army following the end of the Cold War, symbolized by the fall of the Berlin War in 1989. To maintain cadet leadership opportunities, the four I companies were not deactivated until 1998. Following the terrorist attacks on September 11, 2001, the Corps was again increased to 4,400 beginning with academic year 2003–04. However, the four I companies were not reactivated until May 20, 2011, to be fully manned for the academic year 2011–12 beginning in August 2011.

Companies are similar to fraternities at American colleges. Cadets room with fellow classmates and eat at tables in the mess hall organized by company, with ten cadets from the same company eating at a table. The mess hall allows for the entire Corps of Cadets to eat at one sitting.

The mess hall, named Washington Hall, was completed in 1926 with three wings. It was expanded to six wings to accommodate the increased Corps in the 1960s. One mess hall wing wall is painted with a mural in egg tempera by Thomas Johnson. He completed the mural in 1936 as part of the Public Works of Art project during the Great Depression. Twenty decisive battles of history and twenty generals are portrayed in the work, which is called "Panorama of Military History." The Class of 1976 sponsored a restoration of the mural in 2006.

In addition to living and eating together, cadets play intramural sports on company teams that compete with other companies for championships. Cadets also march together by company in parades.

A mural depicting famous historical battles and figures dominates one wall of a wing in the mess hall.

A cadet from each company carries a guidon, or small flag, that leads or guides the company during parades. Companies also have unique patches sewn on their black jackets and nicknames.

Thayer assigned cadets to companies for all four years based on height in order to present a uniform appearance for parades. The tallest cadets were assigned to 1st Company and the shortest to 2nd Company. When the four companies were formed, the tallest went to the A and D companies, which were nicknamed the "flankers." The shortest cadets went to the center two companies, B and C, earning the nickname "runts."

Starting in 1957, cadets were assigned to companies based on scholastic abilities, physical fitness, and varsity athletic participation to ensure an equal distribution of talent. Height was no longer a consideration. Cadets are now assigned to companies to promote diversity among the Corps of Cadets.

Henry Flipper was the first African American to graduate from the Academy, in 1877. (Special Collections, West Point Library)

Diversity also included more consideration of minorities. The first African-American graduate, Henry Flipper (Class of 1877), endured ostracism at West Point, reflecting the segregation of African Americans in the United States that lasted into the 1960s. Congress passed Civil Rights legislation in 1964, which spurred West Point to increase minority admissions so that the Academy and the Army more accurately mirrored American society.

Women were first admitted to West Point by an act of Congress in July 1976. One hundred nineteen started and sixty-two graduated as part of the Class of 1980. Initially, women were assigned to only the first company in each battalion or twelve companies to allow enough women in a company to room together and support each other.

Presently, women are assigned to every company and comprise about 15 percent of the Corps.

Diversity at the Academy goes beyond race or gender. Geographic diversity has been present at West Point since 1828, when Secretary of War James Barbour requested each member of Congress to nominate a candidate from his district for appointment to the Academy. Congress enacted this system into law in 1843, with one cadet authorized for each member of the House of Representatives plus nominees for territories and the District of Columbia. The President was allowed ten "at large" appointments, usually for sons of Army officers. A modified system exists to this day, with each Congressional district allocated five, each Senator and the Vice President also allowed five nominees, and three hundred appointments for the President.

In 1962, cadets were "scrambled" or reassigned to different companies after their first two years. This policy is currently used as it strengthens the cadet chain of command for the senior and junior cadets. The First Class cadets (seniors) are the officers of a company, which includes a company commander, executive officer, and four platoon

The first women cadets cheer on the Army team during fall 1976. (Author's Collection, Photo by Eilene Harkless Moore)

leaders. This chain of command is similar to that of an Army company, and the cadets rotate leadership positions for each semester in order to maximize leadership opportunities. First Class cadets are designated by a class shield consisting of the helmet of Athena on a black background.

Second Class cadets (juniors) form the noncommissioned officer ranks, with a first sergeant, four platoon sergeants, and four sergeants per platoon who perform as squad leaders. Their shield is the helmet of Athena on a gray background. Third Class cadets (sophomores) are corporals and serve as team leaders to the Fourth Class cadets (freshmen). The Third Class shield is Athena's helmet on a gold background. Fourth Class cadets have no shield.

In 1817, Thayer devised a system of chevrons or stripes to designate cadet officers and NCOs for the companies. A modified version of Thayer's chevron system of rank is used to this day. Four stripes designate a cadet captain company commander. He works closely with the legal company commander, the tactical officer, whose office is in the cadet company area. There is also a TAC NCO who provides an important role model for cadets in interacting with NCOs. Working well with NCOs is crucial for the cadet with a future as a lieutenant in the Army.

West Point cadets may spend a semester away from their company at another academy. Exchange cadets from the Air Force, Naval, and

Present-day women cadet cheerleaders cheer on West Pointers during a sporting event.

Coast Guard Academies, along with the Royal Military College of Canada, participate in the Service Academy Exchange Program. Cadets who are juniors study at West Point for a semester. They wear their own academy uniform for parades.

International cadets spend all four years at West Point, wear cadet uniforms, and receive a degree from the Academy. After graduation, they are commissioned in their own nation's army. The first two international cadets at West Point, just twelve and fourteen years old, were from Chile and arrived in 1816. Their poor English-speaking skills, however, precluded them from graduating, and they eventually returned to Chile from West Point. Antonio Barrios from Guatemala was the first graduate from another country, graduating in 1889. Congress authorized admission of foreign cadets based on American international relations. Following the acquisition of the Philippines after the Spanish–American War in 1898, cadets were allocated from that country for many years. Latin American countries also were allocated spaces for cadets. Following the end of the Cold War in the early 1990s, many graduates were from Eastern European countries such as Poland, Romania, and Croatia. A recent graduate was from Afghanistan. Twenty-four international cadets are allowed to attend West Point each year. They are completely integrated into companies and wear the cadet uniforms.

The helmet of Athena is used for the cadet shoulder patch (left bottom) worn on the left shoulder of the black jacket and the subdued version worn on the ACU uniform (right bottom). The class shields (top) also depict her helmet. (Author's Collection, Photo by Eilene Harkless Moore)

Company G-3: The Gophers

Company G-3, nicknamed the "Gophers," provides an example of the thirty-two companies comprising the Corps of Cadets and a snapshot of a company during academic year 2010–11. G-3 was founded in 1967 as part of the expansion of the Corps to 4,400 cadets and from two to four regiments. G-3 was one of eight companies during 2010–11 in the 3rd Regiment, nicknamed "the Wolfpack." The regiment was activated on July 1, 1965. I-3 was deactivated in 1998 and reactivated on May 20, 2011, for academic year 2011–12

Major Robert Bonner, the company tactical officer, is unique among the TACs as he is the Air Force exchange officer. West Point and the United States Air Force Academy exchange TACs for a two-year tour at each other's academy. Major Bonner served as an Air Officer Commanding (equivalent to a TAC) at the Air Force Academy prior to arriving at West Point in May 2010. He is an Air Force Academy graduate of the Class of 1997 and has served as a communications officer supporting various joint and coalition units on multiple deployments overseas. He also earned a Master of Arts degree in Counseling and Leadership from the University of Colorado as part of his training to be a TAC.

Major Bonner is assisted by TAC NCO Sergeant First Class Justin Schreppel. Like most Army officers and NCOs assigned to West Point, SFC Schreppel is a veteran of numerous deployments to Iraq and Afghanistan. Bonner focuses on working with the cadet officers of the company, the First Class and the Fourth Class who are privates, while

BOTTOM LEFT: Major Bonner, U.S. Air Force, the G-3 tactical officer (left), confers with Cadet Captain Paul Erickson from the Class of 2011 (right), the company commander.

BOTTOM RIGHT: Sergeant First Class Schreppel (left) instructs Cadet Rita Snyder (Class of 2013) on counseling techniques.

SFC Schreppel works with the cadet NCOs, the Second and Third Class cadets. Their offices are located among the cadet rooms in the cadet company area to facilitate interaction with the cadets.

Bonner is responsible for the development of 145 cadets, with 33 First Class, 35 Second Class, 38 Third Class, and 39 Fourth Class cadets. Two cadets are from other countries: one from Columbia, and one from Latvia. They wear the same uniforms as the other cadets because they are attending for all four years. He commands the company through a cadet chain of command that mirrors that of an Army company, with a cadet captain company commander, cadet lieutenant executive officer, and four cadet lieutenant platoon leaders. Bonner meets daily with the cadet company commander and weekly with the cadet company staff to coordinate company activities and address administrative actions.

On the NCO level, there is a cadet company first sergeant, who is a Second Class cadet, four cadet platoon sergeants, and sixteen squad leaders (four per platoon). Schreppel instructs cadets on how to conduct counseling to improve cadet performance of military duties.

Cadets are in G-3 for two years. The Fourth and Third Class cadets are reassigned, or "scrambled," after two years into different companies to allow for more leader development opportunities. The Second and First Class cadets spend their last two years together as the leaders of the company. To maximize leadership opportunities, the positions are rotated at the end of the first semester so there are two cadet company commanders of G-3 for the academic year.

Just as each company has a TAC and a cadet company commander, each regiment also has a Regimental Tactical Officer and a cadet regimental commander. The RTO for academic year 2010–11 was Lieutenant Colonel John Vermeesch. He is a Class of 1990 West Point graduate and served as a company TAC in a previous tour at West Point. He earned a Master of Science Degree in Counseling and Leader Development from Long Island University, C.W. Post, as part of his training to be a TAC. He is a combat veteran with two tours in Iraq as an infantry officer. He returned to West Point after a successful battalion command with the 1st Infantry Division at Fort Riley, Kansas, and serves as a senior mentor and role model for the cadets of the 3rd Regiment.

Lieutenant Colonel John Vermeesch (center), 3rd Regimental tactical officer, confers with Cadet Jae Yu (Class of 2011) and Major Robert Bonner, U.S. Air Force.

CHAPTER 9

Pass in Review

Parades from the Plain to 1600 Pennsylvania Avenue

Cadets have marched upon the Plain in reviews or parades since the founding of the Academy in 1802. The small size of the Corps of Cadets prior to the organization into companies in 1812 prevented the large-scale parades so familiar to present-day visitors. Superintendent Alden Partridge started not only the legacy of the cadet gray uniform, but also the iconic image of the uniform on parade for an admiring crowd. Superintendent Sylvanus Thayer built upon this image with cadet marches to various cities in the early 1820s.

The most significant march of cadets was to Boston from July 20 to August 26 in 1821. The first flags, called "colors," were presented to the Corps of Cadets by the admiring citizens of Boston on August 11, 1821. A national standard and battalion color composed this first set of colors. Ever since then, in various designs, the Corps of Cadets have marched with colors. The battalion color featured the Roman goddess Minerva, the Roman version of the Greek goddess Athena, one of the symbols of West Point. The helmet of Athena is featured on the West Point crest, shoulder patch, and class shields. Athena was adopted as the patron of West Point because she is the symbol of wisdom and knowledge of the arts of war. Various statues of Athena are scattered throughout buildings at West Point.

The national standard (flag) was an eagle on a dark blue background. This followed the Army custom of a national color with an eagle and a regimental color. Two colors were carried until 1841,

A Cadet Company on parade on the Plain.

The Greek goddess Athena is a traditional symbol of knowledge.

when this was changed to just one flag. In 1870, the Stars and Stripes became the Corps color with the inscription "U.S. Corps of Cadets" written on the center stripe. In May 1894, a Corps color of an eagle with a blue background was authorized along with the national color of the Stars and Stripes. In July 1902, the present-day gray color with the West Point crest was first used along with the Stars and Stripes.

A cadet color captain is in charge of the color guard. He is assisted by a color lieutenant as the executive officer. They supervise two teams of six cadets. Each team is composed of a color lieutenant and five color sergeants. The two teams stay busy throughout the year representing the Corps of Cadets at various ceremonies and functions. The normal composition of the color party for a review is five cadets. Two color sergeants march as the escort to the three colors. The colors presently carried are the Stars and Stripes, the Army Flag, and the Corps colors. At events outside of West Point, the color guard also will carry the Academy flag to represent the entire Academy, which forms a six-cadet color party.

General Officer flags are carried by color corporals who are Third Class (sophomore) cadets in training for becoming color sergeants as Second Class (junior) cadets. The color captain supervises the employment of GO flags just prior to the start of a review. Each general present at a review is entitled to a red flag with stars on the flag matching the number of stars for the general. The superintendent as a lieutenant general rates three stars, while the commandant and dean as brigadier generals fly one-star flags.

For academic year 2010–11, the cadet color captain was Benjamin Clark (Class of 2011) and the executive officer was his twin brother,

A Stars and Stripes with "U. S. Corps of Cadets" written on the center red stripe from the mid-1890s. (West Point Museum, Photo by Eilene Harkless Moore)

Zachariah Clark. They supervised the color guard's marching in eleven reviews or parades, starting with the Acceptance Day parade on August 14, 2010, and ending with the Graduation Parade on May 20, 2011. The color guard also presented the colors at more than one hundred

Cadet colors from the late 1890s. The background originally was dark blue but has faded over time. (West Point Museum, Photo by Eilene Harkless Moore)

105

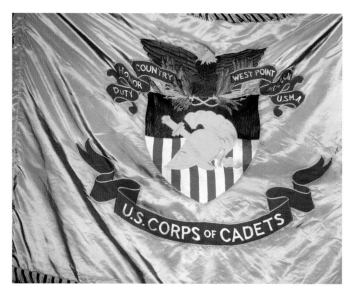

events representing the Academy, ranging from a football game at Yankee Stadium against Notre Dame to the Special Olympics at West Point.

Acceptance Day marks the beginning of the academic year and is the first parade for the entire Corps of Cadets. The new cadets, who have completed the six-week Cadet Basic Training nicknamed "Beast Barracks," are formally accepted into the Corps of Cadets as Fourth Class cadets, or "plebes." The plebes form up by academic year company formations and then, after acceptance into the Corps, march to the rear of their new company, which will be their home for the next two years. The entire Corps of Cadets then pass in review for the superintendent and the proud parents of the plebes.

TOP LEFT: General officer flags flown at a review for the superintendent with three stars (left) and commandant with one star (right).

TOP RIGHT: Present-day Cadet Colors feature the Academy crest embroidered on silk.

Reviews are conducted for each home football game, with just two regiments marching per football review. The entire Corps of Cadets participated in the Thayer Award Review held on October 7, 2010. This award, given for distinguished public service by an American, is made by the superintendent in conjunction with the West Point Association of Graduates, the Academy's alumni association. The 2010 Thayer Award was presented to James Baker, former secretary of state and secretary of the treasury. He inspected the cadets standing up while riding in a quarter-ton utility truck, better known by its World War II nickname, "the Jeep," alongside Lieutenant General David Huntoon, Jr., the Academy's fifty-eighth superintendent.

The Color Guard with all four colors including the Academy flag (third from left) marches in the New York City Veteran's Day parade on November 11, 2010, in full dress wearing gray trousers for winter. The band follows the Color Guard, also wearing their winter uniform with blue trousers. (Photo by Staff Sergeant Crissy Clark, West Point Band)

For academic year 2010–11, the Corps of Cadets was organized into a brigade of four regiments; each regiment was composed of two battalions with four companies in each battalion. When arrayed on the Plain, there are two lines of sixteen companies, one behind the other. Standing in front of the first line of companies are the eight battalion staffs. In front of them are the four regimental staffs, and centered in

Cadet Color Captain Benjamin Clark leads the General Officer flag bearers for deployment at a review.

PASS IN REVIEW

Former Secretary of State James Baker (standing in Jeep, right) inspects the Corps of Cadets riding in a Jeep escorted by the superintendent (left rear standing) and first captain (left front standing).

front of all is the brigade staff. Each staff consists of cadet officers who assist the commander. Personnel, supply, training, and various other staff officers complete the picture.

First Captain Marc Beaudoin is the cadet commander of the entire Corps. He is assisted by a brigade staff. Notable first captains of the past include John Pershing (Class of 1886), commander of the American Forces in Europe during World War I, and Douglas McArthur (Class of 1903), commander of American Forces in the Southwest Pacific during World War II and of American Forces in Korea.

Centered with the first line of companies, behind the brigade staff is the color guard. The positioning of the color guard dates back to the

ABOVE: First Captain Marc Beaudoin (Class of 2011) salutes (center) the passing colors with his staff formed up behind him.

BELOW: The long gray line of companies is assembled, ready for inspection with the colors centered among the companies.

linear tactics used during the early 1800s by Napoleon. Regiments placed their colors in the center of a line of ten companies. Soldiers standing shoulder to shoulder fired musket volleys and used the colors as a reference point and a visible symbol to rally around the colors during the smoke filled confusion of battle. Over time the linear gray-coated cadets were described as "The Long Gray Line."

Through the years, cadets have marched in many different places for notable events. The most important event for a cadet to march in is a Presidential inauguration. The first time cadets marched in an inauguration was for the second inauguration of Ulysses Grant, the first West Point graduate to be elected President of the United States. On March 4, 1873, the entire Corps of Cadets marched down Pennsylvania Avenue. Since then, cadets have marched in every President's inaugural parade. Dwight Eisenhower (Class of 1915) was the second graduate elected President, and he served two terms, from 1953 to 1961.

In recent inaugural parades, because of the academic schedule, the entire Corps does not participate. Instead, cadets volunteer to form a composite company to march in parades. The brigade staff lead the cadets, but the first captain marches with the Army staff. The Corps of

The entire corps marched in President Harry Truman's inaugural parade in January 1949, wearing long overcoats. (Author's Collection, Photo by Eilene Harkless Moore)

Cadets leads the entire inaugural parade as the order of precedence for military units marching in the parade places the West Point cadets ahead of all other military units.

Cadets also participate in more somber events such as funerals of distinguished graduates. The Class of 1951 marched in the India white uniform for the funeral of General John Pershing on July 19, 1948, in Washington, D.C. He was buried at Arlington National Cemetery.

During the summer, cadets march in the white over gray under arms uniform. The most familiar uniform, full dress over white under arms is worn for early fall and late spring reviews. From October through April, full dress gray with gray trousers are worn.

The Class of 1951 represented West Point at General John J. Pershing's funeral in July 1948, wearing the India white uniform. (Author's Collection, Photo by Eilene Harkless Moore)

On Parade with G-3

The Gophers were part of the first spring review held on April 30, 2011, with the 3rd and 4th Regiments. Two practices were conducted during the week prior to the event to march out the winter blues. For practice reviews, the class uniform with full dress hat, white gloves, and rifle is worn. This uniform is rarely seen by the public. Along with the TAC, evaluators with clip boards closely observe and perform on-the-spot corrections as needed to drill the cadets back into sharp marching form.

On Saturday, the cadets start their preparation at 0945 hours, with the first calling of the minutes by plebes. "Minute calling" is a time-honored duty performed by plebes who stand in the company hallways and shout out to the upperclass cadets how many minutes are left until formation. The minute caller also tells everyone of the proper uniform: full dress over white under arms. This informs the other cadets that they are wearing their full dress coats with white trousers. "Under arms" means wearing white gloves, waist belts, and white cross belts carrying the M-14 rifle with fixed bayonet. They wear the full dress hat with an eight-inch black wool pompon. Cadet officers wear a red sash and carry a saber instead of a rifle. Those First Class cadets in leadership positions, such as the company commander and platoon leaders, wear a taller plume made of black cock's feathers. Those First Class cadets not in leadership positions march as the rear rank and wear the pompon to fit in with the marching rifle ranks.

The company forms up behind a passage way through the barracks called a "sally port" an hour before the start of the parade. Only fifty-five cadets are required out of the 145 assigned to the company to

Major Bonner (left) observes G-3 present arms during a practice.

Cadets practice (center front) and wait at a sally port entrance prior to the start of a review.

march. This is to facilitate those cadets who cannot be present due to conflicting scheduled activities such as intercollegiate sports competitions, academic trips, or various other assigned support duties. For this review, the executive officer, substituting for the company commander, leads with the guidon (company flag) and then four platoon leaders. The main body of the company is composed of five ranks (rows) of rifles with eight in each rank, with one rank of sabers and the first sergeant bringing up the rear.

Companies form up by height, with the tallest cadets in the front starting from the right and the shortest cadets in the rear. The tallest

BOTTOM LEFT: Two plebes prepare to call the minutes in a G-3 hallway prior to a review.

BOTTOM RIGHT: Air Force Cadet Dakota Newton (right), wearing his parade dress uniform, confers with other cadets before review begins.

Staff Sergeant Daniel Pierce (left) and Brigade Adjutant Johnny Garcia (Class of 2011) (right) march onto the Plain.

cadet ends up as the right guide, ensuring that the company marches to the correct place among the other sixteen companies on the Plain.

For this review, the tallest cadet in the company happened to be the Air Force Academy exchange cadet, Dakota Newton, a 2 degree (which is Air

The Hellcats march onto the Plain under the gaze of Dwight Eisenhower (rear) (Class of 1915).

Bugles play to alert the cadets that the review begins in two minutes.

Force Academy name for a junior). The parade uniform worn by Newton was designed in 1954 by Academy Award-winning Hollywood film producer and director Cecil B. DeMille for the then newly established Air Force Academy. This distinctive blue uniform stands out in the field of gray.

Drums and bugles give the final warning to the cadets.

While G-3, along with the other fifteen cadet companies, are preparing out of sight of the spectators, the West Point band and brigade staff take positions on the Plain to orchestrate the review.

First on the Plain are the brigade adjutant and a trombone player, followed closely behind

George Washington (top) seems to urge forward the surge of cadet companies from the sally ports.

by the Hellcats. The rest of the band joins the Hellcats on the parade field to the far right of the line using the single trombone player as their guide.

The adjutant starts the review by ordering the drum major to "sound attention." The bugle's blare signal to the cadet companies a two-minute warning before they start to march. Next, the adjutant gives the command to "sound adjutant's call." The sound of the drums and bugles are the final warning to the waiting cadets. Hand-held radios are used by cadets stationed in the sally ports to precisely synchronize the initial companies march out of the sally ports. The band plays music as the companies move through the sally ports onto the Plain. A steady drumbeat is the key to keeping the marching cadets all in step with

Members of the Class of 1951 move onto the Plain for the review.

each other. Once the companies are formed up on the Plain, the brigade staff marches out from the bleacher stands headed by the first captain to join the adjutant in front of the two regiments.

For the spring 2011 review, two special groups were honored. The Class of 1951 was present for

their sixtieth reunion. A quarter of this class was commissioned into the Air Force because there was not an Air Force Academy until 1954; the first Air Force Academy class graduated in 1959. Army retirees form the second group honored by this review. Every year retirees are recognized for their service to the nation during a retiree open house held at all Army posts. A fortunate group of Army officers and civilians retire this day in front of the Corps of Cadets.

G-3 presents arms (salutes with their rifles and sabers) to the reviewing party and for the playing of the national anthem. After the appropriate honors have been rendered, the order is given to "pass in review." As G-3 marches in front of the reviewing party they salute

Soldiers and civilians at their retirement ceremony salute during the playing of the national anthem.

G-3 presents arms.

G-3 passes in review.

them by conducting an "eyes right" where all in the company turn their heads to the right as they pass in front of the superintendent and other members of the reviewing party. Those cadets in the closest ranks to the reviewing party do not turn their heads but march straight ahead to keep the rest of the company straight on line. As they continue marching forward, G-3 completes the review by returning through the same sally port behind the barracks where they fall out and return to their rooms.

G-3 (center background) returns to their barracks as the Brigade staff (foreground) looks on.

ABOVE: The marching ranks of G-3 conduct "eyes right."
BELOW: The superintendent returns the salute of G-3.

CHAPTER 10

The Ring

Symbol for a Lifetime

The class ring is more important to a West Point graduate than any other item. The Class of 1835 started the tradition of wearing class rings, the first at any American university. Each class, except the Class of 1836, has worn this symbol of their four years at West Point, a symbol that lasts a lifetime.

A Class of 1835 ring does not exist. Only a wax impression of the class crest survives. This crest is a shield surrounded by a banner inscribed with the class motto in Latin "Amicus Periculique Foedus" (Danger Brings Forth Friendship). The crest was part of the bezel or front of the ring where the present-day stone is placed.

The earliest enduring ring is from the Class of 1837. This ring is smaller than present-day versions as nineteenth-century fashion called for men's rings to be worn on the little finger. In 1869, a Cadet Ring Committee was formed to design a ring. The sides of a ring, called the shank, are inscribed with the class crest on one side and the West Point crest with motto on the other. This design practice started with the Centennial Class of 1902. The class motto was replaced by the Academy motto of "Duty, Honor, Country" on the ring. The class motto did not reappear on the ring until the Class of 1970 decided to incorporate the class motto into the class crest design. Certain design elements must appear in a class crest, including the American eagle, class year, class motto, cadet saber, and the letters "USMA." The Class of 1976 crest, designed to commemorate the bicentennial of the United States,

G-3 stands at attention on the Plain. During their tenure at West Point, each will wear a class ring, a tradition that dates back to 1835.

includes a minuteman soldier in lieu of the letters USMA in the design and incorporates an appropriate class motto: "Spirit of 76." The exception to policy to delete the letters USMA was granted because of the American Bicentennial.

President John F. Kennedy was accorded a unique honor by the Class of 1962. He was presented a class ring with the Presidential seal engraved on the bezel when he spoke at the class graduation on June 6, 1962. The ring was reacquired by the class and given to West Point at the class's forty-fifth reunion in October 2007.

Cadets are presented their ring during a ceremony held at the beginning of their First Class year. The uniform worn for the ring presentation is the India white with red sash. Ring weekend is one of the highlights of a cadet's life at West Point. A formal dance or hop is still held for this special time in a cadet's life at West Point.

Embroidered class crest of the Class of 1977 with the motto "Esprit de Corps." (Author's Collection, Photo by Eilene Harkless Moore)

Class crest engraved on one shank of 1977 class ring. (Author's Collection, Photo by Eilene Harkless Moore)

The West Point crest is engraved on the shank of all class rings. (Author's Collection, Photo by Eilene Harkless Moore)

Class of 1962 Ring

This ring, with the Presidential Seal in place of a stone, was presented to President John F. Kennedy at the Graduation of the Class of 1962 on 6 June 1962.

The ring was reacquired by the Class of 1962 and presented to West Point at the 45th Reunion of the Class in October 2007.

President John F. Kennedy's ring presented to him by the Class of 1962 with the Presidential seal engraved on the bezel. (West Point Library, Photo by Eilene Harkless Moore)

A TAC (left) presents to a member of the Class of 2011 his class ring. The cadet wears the India white uniform. (West Point Public Affairs Office)

The ring dance traditionally includes a replica of the class ring, as shown by this couple from the Class of 1951. (Author's Collection, Photo by Eilene Harkless Moore)

BOTTOM LEFT: Unique class crest of 1976 substituted a minute man soldier in lieu of the letters "USMA" to acknowledge the American bicentennial. (Author's Collection, Photo by Eilene Harkless Moore)

BOTTOM RIGHT: Class crest of the Class of 2011. (Author's Collection, Photo by Eilene Harkless Moore)

A class ring from the Class of 2011 is shown belonging to Stanley Gorzelnik, a G-3 Gopher. The class crest incorporates the traditional design elements along with the American flag and the statue of George Washington. Reflecting the realities facing graduating cadets since the attacks on September 11, 2011, their motto is "For Freedom We Fight."

A new tradition started by the West Point Bicentennial Class of 2002 is the Ring Melt. Prior to the making of the rings, donated class rings, usually of deceased graduates, are melted down and the gold is used to cast the current class rings. This program is sponsored by the West Point Association of Graduates as way to connect present cadets with those who have gone before them.

CHAPTER 11

The 21st Century Cadet

Cadet Matthew Fiorelli (Class of 2014) is an example of a twenty-first-century cadet. He entered West Point on June 28, 2010, along with 1,385 other new cadets on "R-Day" (Reception Day), including 250 women (18 percent), 131 Asian Americans (9.5 percent), 126 African Americans (9.1 percent), and 125 Hispanic Americans (9 percent).

Fiorelli formed up with his fellow new cadets on the Plain for the oath after spending a busy first day in processing at West Point. On their first day, incoming new cadets start reporting in at 6:30 a.m. and then at thirty-minute intervals until all have reported by 9:30 a.m. to the cadet in the red sash. Their first uniform is black shorts, black socks, black shoes, and white t-shirts. They spend the rest of the day being issued uniform items, learning to salute, and learning to march. At 6:25 p.m., eight companies, lettered A–H, march out onto the Plain with new cadet Fiorelli among the ranks of the 3rd platoon, G Company. His assignment to this platoon means he will become a G-3 Gopher for this academic year.

The cadet cadre lead and stand out front in their dazzling India white uniforms, the officers wearing red sashes. The new cadets are in white over gray (white shirts with gray trousers) with white gloves and no hats. Proud parents watch their sons and daughters take the cadet oath.

Six weeks later, on August 14, Fiorelli, having endured Cadet Basic Training, is formally welcomed into G-3 during the Acceptance Day

Cadet companies at parade rest during an event on the Plain

parade. He marches for the first time as a cadet and hears the first of many future commands: "Pass in Review." The Acceptance Day Parade is different from other reviews. At the beginning of the ceremony, the new cadets form apart from the rest of the Corps. They stand in company formations directly in front of the bleachers. The three upper classes march onto the Plain through the sally ports. The Class of 2014 is presented to the Corps of Cadets and marches forward in two waves to join their companies. The newly formed companies with all four classes pass in review, and Cadet Fiorelli marches for the first time as a full-fledged member of the Corps.

He and his fellow cadets begin classes the following week in diverse subjects such as chemistry, English composition, history, psychology, literature, calculus, and—every plebe's most memorable course—boxing, which is part of the required physical education classes.

His day starts at 6:30 a.m. with reveille, then breakfast formation at 6:55 a.m. His first class is at 7:30 a.m. Four class periods of fifty-five minutes each are offered in the morning, and two class periods in the afternoon. Lunch formation is held at 12:05 p.m. with the meal from 12:10–12:35 p.m. Company athletics, drill, and military and physical training are scheduled from 4:15–5:45 p.m. Dinner is only mandatory for all cadets on Thursdays, starting at 6:45 p.m. with optional "grab and go" food available the other days. Fiorelli's day ends with Taps at 11:30 p.m. and lights out at midnight.

New cadets march onto the Plain wearing their second uniform of white over gray with the cadre in India white. (West Point Public Affairs Office)

Cadet Matthew Fiorelli studies in his room using his newly issued laptop computer.

All cadets must take twenty-six core courses, an information technology course, and three engineering course. The Math–Science–Engineering sequence consists of four semesters of mathematics, two of physics, two of chemistry, and at least three of engineering. To graduate, all cadets must complete the thirty required courses and at least ten electives in their major. Eight military science classes and a rigorous program of physical education also are graduation requirements.

As are his classmates, Fiorelli is issued a laptop computer, which he takes to class with him and uses in his room. He wears the class uniform, which, in the relaxed fashion of the twenty-first century, is a

Cadet Matthew Fiorelli wearing the "As For Class" uniform walks through the G-3 company area on his way to class.

ABOVE: Regardless of the century, shoes still must be shined.

RIGHT: Hats, uniforms, and footwear are required to be displayed in the closet.

short-sleeved shirt. A long-sleeved shirt with tie, worn in winter, is now optional. Although uniforms have changed over the years, some things, such as keeping shoes shined, have not changed since the nineteenth century. Uniforms, for instance, still must be correctly arranged in the cadet's room closet.

During his Third Class year, Fiorelli will declare a major from more than forty that are offered at West Point. Just less than half of cadets

Cadets in black jackets return from class.

chose a math–science–engineering major. In a nod to Thayer, all cadets must take at least a three-course engineering track in one of seven areas ranging from Thayer's civil engineering to nuclear and environmental engineering.

Cadets take an introduction to philosophy course, usually during their Third Class year. This course was launched during the 1980s and

A cadet (right) salutes and reports to Lieutenant Colonel Jeffrey Wilson that the section is present and ready to receive instruction in philosophy class.

Lieutenant Colonel Dave Jones (Class of 1985) presents (left) at the "Inspiration to Serve" event as he tells of the sacrifice made by a fallen graduate fighting in Iraq to future officers of the Army. (Courtesy of Simon Center for the Professional Military Ethic)

is one of four required courses from the Department of English and Philosophy. Cadets spend almost half the semester on the Western ethical tradition of Just War theory. Lieutenant Colonel Jeffrey Wilson is one of the professors who teach this required course in philosophy. He is a Reserve Officer Training Corps graduate and holds a Masters degree in philosophy and a Doctorate in Education. He served a tour in Afghanistan and brings real-world experience on the rules of engagement designed to protect civilians, versus killing an enemy who uses civilians

RIGHT: Lieutenant Colonel Wilson makes a philosophical point as a cadet makes notes on her issued laptop computer.

FAR RIGHT: Neil Kanneberg holds a panda at the Chengdu Panda Breeding and Research center. (Courtesy of Neil Kanneberg)

as human shields. Applying Just War theory to this dilemma sparked spirited discussion in the classroom. In 2011, Wilson was one of approximately half the faculty who are not West Point graduates. About 21 percent of the faculty is civilians; the remaining are nongraduate commissioned officers.

As a Third Class cadet, Fiorelli will participate in the "Inspiration to Serve" tour of the West Point cemetery. This event is the capstone of the Third Class Professional Military Ethic Education Program. The purpose is to acquaint cadets with the ultimate sacrifice paid by graduates. Volunteer presenters, typically former classmates of the fallen graduates, tell the story of their now-silent friends. Many of the fallen are twenty-first century graduates who were killed while fulfilling their duty to country fighting the Global War on Terrorism in Iraq or Afghanistan.

G-3 cadet Neil Kanneberg (Class of 2012) standing in Tiananmen Square in Beijing, China. (Courtesy of Neil Kanneberg)

The Simon Center for the Professional Military Ethic (SCPME) was founded in 1998 as a center of excellence to promote cadet character development and foster the concept of "officership" within each cadet. William E. Simon, a patriotic businessman, former secretary of the treasury, and generous philanthropist, provided funding to establish the center.

Cadets receive instruction on the Honor Code through programs run by the Cadet Honor Committee. The SCPME provides support for honor education with textbooks and other materials. Fourth Class through Second Class cadets receive instruction through the Professional Military Ethic Education Program. The curriculum complements other Cadet education in values, officership, and leader skill development. The National Conference on Ethics in America is sponsored by the center each fall. The three-day conference involves cadets, students from more than sixty academic institutions, and leaders from business and industry.

A Battle Command Conference is held each year in April to expose First Class cadets to junior officers and noncommissioned officers recently returned from combat in Iraq or Afghanistan. Cadets meet in small groups to discuss leadership during war.

Officership MX 400 is the capstone course for First Class cadets. Cadets take this course during their senior year to apply their forty-

Matthew Mitchell (Class of 2011) (right) gives pointers on handling an Army weapon to a scout during the 2011 camporee.

seven month West Point experience toward their future as officers in the Army.

Another G-3 cadet, Second Class Cadet Neil Kanneberg (Class of 2012), demonstrates the worldwide opportunities for study available to the twenty-first century cadet. He spent the second semester of academic year 2010–11 at Peking University in Beijing, China, as a participant in the semester abroad program. More than 140 cadets studied at foreign universities during the 2010–11 academic year as part of this program. Kanneberg took Chinese as his required two semesters of foreign language study. He double majored in Chinese and economics. The classes taught in Mandarin Chinese are in grammar, culture, media, and oral Chinese language, and all are credited to his major.

A cadet (left) instructs scouts on how to march during the camporee.

The graduating class forms the reviewing party for the Graduation Parade and salute for the final time for a Pass in Review. (West Point Public Affairs Office)

Admiral Mike Mullen, chairman of the Joint Chiefs of Staff, speaks at the 2011 commencement. (West Point Public Affairs Office)

First Class Cadet Matthew Mitchell was one of the Gophers who missed the April 30, 2011, review because he was the cadet in charge of the largest annual Boy Scout Camporee held in America. Cadet Mitchell, who is also an Eagle Scout, supervised 225 cadets, including 23 from G-3, in supporting 5,575 scouts at the 49th West Point Boy Scout Camporee held from April 29 to May 1, 2011. This is the largest community outreach program conducted by cadets.

Cadet Mitchell marched in his last parade as a cadet during the Graduation Parade held on May 20, 2011. This parade, like the Acceptance Day parade, is unique. All First Class cadets march out at

The Class of 2011 toss their hats to celebrate the end of their cadet days. (West Point Public Affairs Office)

the front of their companies. They then form up by company groups on the Plain in front of the bleachers, similar to how they stood as new cadets almost four years previously. The soon-to-be graduates remove their dress hats as the Corps of Cadets conducts the last "Pass in Review" for the graduating class. The West Point Alma Mater song stanza "may that line of gray increase from day to day," takes on new meaning to these newest members of the "Long Gray Line."

Navy Admiral Mike Mullen, the chairman of the Joint Chiefs of Staff, gave the academy's commencement address on May 21, 2011. He spoke about the need to work alongside civilians to win wars and reiterated that soldiers are also part of the greater American community. The chairman also talked about how officers should strive to be soldiers as well as statesmen. Mullen's sentiments are brought to life in the community efforts of graduates such as Mitchell and in the studies undertaken in semester-abroad programs such as the one Kanneberg participated in.

As part of the traditional hat toss that ends the graduation ceremony, 1,031 graduates threw their white caps up into the air. This amazing rain of white hats signals that the Class of 2011 has joined more than 67,000 previous graduates of the Long Gray Line.

The Combating Terrorism Center

The Combating Terrorism Center (CTC) was founded at West Point in February 2003 through the vision and financial support of Vincent Viola (Class of 1977). He was chairman of the New York Mercantile Exchange in 2001 and personally witnessed the attack on the Twin Towers in New York City while driving into work on that tragic morning of September 11, 2001. He was spurred into action by the terrorist attacks and spearheaded a funding drive by the Class of 1977 to provide primary financial sponsorship for the center. He continues to provide leadership and commitment to the CTC to allow for the continued study of the terrorist threat.

Cadets may receive a minor in terrorism studies, the only one of its kind available in the United States. This minor program includes five courses taken in addition to the required core course in international relations. All cadets receive a four-class block of instruction on terrorism as part of their international relations course.

The influence of the CTC extends well beyond West Point. The Center provides more counterterrorism education to federal, state, and local governments than any other organization in the United States. For example, the Fire Department of New York City has partnered with the CTC for an eleven-week graduate seminar. The Federal Bureau of Investigation also works with the center to provide FBI new special agents with education on counterterrorism.

BOTTOM LEFT: Cadets Maria Casaletto (left) and Christopher Best (right) with FDNY Chief James Kane (center) participate in a functional exercise with the New York City Fire Department as part of their homeland security course. (Photo courtesy Countering Terrorism Center)

BOTTOM RIGHT: Vincent Viola (Class of 1977), founder of the Combating Terrorism Center. (Author's Collection, Photo by Eilene Harkless Moore)

CHAPTER 12

The West Point Cemetery

Gallery of the Fallen

The West Point Cemetery was officially established in 1817, but soldiers were first buried there, a fourteen-acre promontory overlooking the Hudson River, in 1782. The scenic spot is one of the few level areas at West Point and originally was called "German Flats" because Pennsylvania German-descended soldiers first camped here during the Revolutionary War. The first known soldier buried here is Dominick Trant, born in Ireland and a soldier in the 9th Massachusetts Infantry. He died of a fever while serving in the Revolutionary War, and General George Washington attended his funeral in 1782.

The only other known grave of a Revolutionary War soldier is that of Margaret Corbin, the first woman to receive a pension from the U.S. Government for war-related wounds. Corbin accompanied her husband John when he joined a Pennsylvania Artillery regiment. She replaced him on his cannon crew after he was killed fighting the British at the Battle of Fort Washington, in New York City, on November 16, 1776. She was wounded firing the cannon, captured, and then released by the British. In 1779, Congress granted her a pension for her injuries. She died in 1800 and was first buried outside of West Point. She was reburied in 1926 when the Daughters of the American Revolution built a monument to her at the West Point cemetery.

The oldest monument dedicated to a graduate was built for Lieutenant Colonel Eleazer Wood (Class of 1806). He was killed in action during the War of 1812 at Fort Eire, Ontario, Canada, on

The oldest known headstone in the West Point Cemetery is of Dominick Trant, an American Revolutionary War soldier.

An inscription for Margaret Corbin, the first woman to receive a pension from the U.S. Government for service as a soldier.

BOTTOM LEFT: This monument is dedicated to Margaret Corbin, who was wounded fighting the British during the American Revolutionary War.

BOTTOM RIGHT: In 1818, the oldest monument to a graduate, Eleazer Wood, was dedicated on the Plain and then later moved to the cemetery.

George A. Custer

George A. Custer was a member of the class of June 1861. The class graduated a year early due to the start of the Civil War in April 1861.

While at West Point, Custer acquired a most unmilitary nickname from his fellow cadets—"Fanny"—due to his long, curly hair and peaches-and-cream complexion. Custer distinguished himself as a cadet by graduating last in his class, and he came in first with the most demerits: 726. A fellow cadet wrote of Custer that "… he had more fun, gave his friends more anxiety, walked more tours of extra guard, and came nearer to being dismissed more often than any other cadet I have ever know."

In fact, Custer was tried by general court martial for neglect of duty as a cadet officer of the guard on the very day before his class graduated. He pled guilty to the charges, but his luck held out. Because of the ongoing Civil War, he was merely reprimanded and allowed to graduate.

He later achieved immortality when his luck ran out at the Little Big Horn.

George Custer was killed on June 25, 1876, during the Little Big Horn, where he and his entire command of more than two hundred soldiers died.

Lieutenant Colonel
Eleazer Wood
inscription on
base of monument.

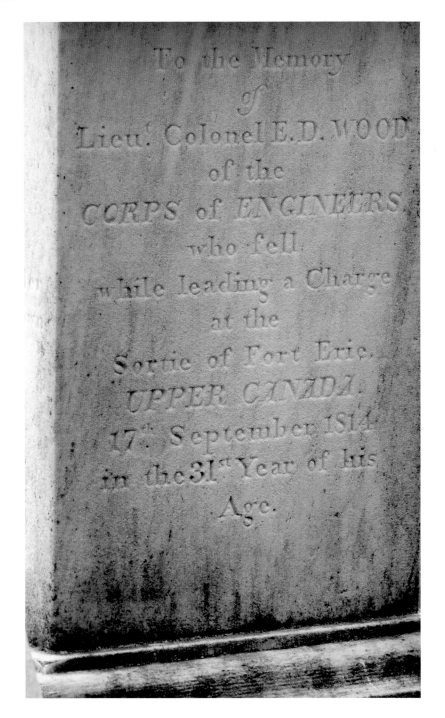

September 17, 1814. Major General Jacob Brown, Wood's commander, was so impressed with Wood's gallantry that Brown had built a fifteen-foot tall stone obelisk monument at his own expense. It originally was located on the Plain and dedicated in October 1818. The monument was

Captain George Derby fought in the War with Mexico and was wounded in action.

This portrait of George Derby was taken while he was on furlough leave in 1844. He wears the cadet furlough uniform of the Army frock coat with no rank. (Special Collections, West Point Library)

then moved to a knoll near the post flag pole and then to the cemetery in 1885, where it stands today.

Wood was one of many West Point graduates who have died in America's wars. The cemetery contains the remains of more than 7,200 people, many of whom are graduates who fought in all the wars of the United States since the War of 1812. From George Derby to Laura Walker, the Gallery of the Fallen honors a graduate from each war (and for manned space exploration), starting from 1846 with the War with Mexico and ending with the ongoing Global War on Terror.

George Derby, Captain, (Class of 1846) fought in the War with Mexico and was wounded at the Battle of Cerro Gordo in April 1847.

ABOVE LEFT:
Cushing's cadet
portrait from class
album. (Special
Collections, West
Point Library)

RIGHT: Alonzo
Cushing fought during
the American Civil War
at Gettysburg and was
killed on July 3, 1863.
He was posthumously
awarded the Medal of
Honor in 2010.

TO LEFT: Cadet Custer's portrait from the 1861 class album. (Special Collections, West Point Library)

ABOVE: Custer's remains were first buried at the Little Big Horn battlefield in 1876 and returned to West Point in 1877.

After the war, he was posted to California as a topographical engineer, where he became well known as a writer of humor and was called the funniest writer in America by Mark Twain. He died in 1861 of a suspected brain tumor.

Alonzo Cushing, First Lieutenant, (Class of June 1861) was killed in action at Gettysburg on July 3, 1863, while defending Cemetery Ridge against Pickett's Charge during the Civil War. He was wounded three times yet refused to leave the battlefield as he commanded his artillery battery. He was belatedly awarded the Medal of Honor in September 2010.

George Custer, Lieutenant Colonel, (Class of June 1861) is most famous for his ill-advised battle at the Little Big Horn on June 25, 1876, during which he and his entire command of more than two hundred soldiers were killed. He was an outstanding cavalry

Dennis Michie died fighting in Cuba in 1898.

Portrait of Cadet Michie. (Special Collections, West Point Library)

commander during the Civil War and promoted to brevet (temporary) brigadier general at age twenty-five. At the end of the war, he assumed the peacetime rank of lieutenant colonel.

Dennis Michie, First Lieutenant, (Class of 1892) died fighting during the Spanish–American War in 1898. He was killed in action during the battle of San Juan Hill on July 1, 1898. Like many an athlete

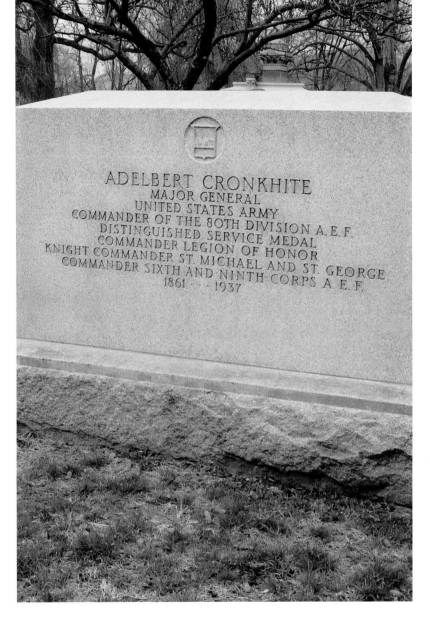

Adelbert Cronkhite commanded a division during World War I.

ADELBERT CRONKHITE
MAJOR GENERAL
UNITED STATES ARMY
COMMANDER OF THE 80TH DIVISION A.E.F.
DISTINGUISHED SERVICE MEDAL
COMMANDER LEGION OF HONOR
KNIGHT COMMANDER ST. MICHAEL AND ST. GEORGE
COMMANDER SIXTH AND NINTH CORPS A.E.F.
1861 · · · 1937

Cadet Cronkhite from the Class of 1882. (Special Collections, West Point Library)

at West Point, he went from the playing field to the battlefield and paid "a soldier's debt."

Adelbert Cronkhite, Major General, (Class of 1882) commanded the 80th Division during World War I. His division fought as part of General Pershing's American Expeditionary Force in France against the Germans.

Norman Cota fought in both World Wars and stormed the beaches at Normandy on June 6, 1944.

Norman Cota's cadet portrait. (Special Collections, West Point Library)

Norman Cota, Major general, (Class of April 1917) stormed the beaches at Normandy on June 6, 1944. He was the first general ashore at Omaha Beach and led soldiers off the beach to attack heavily fortified German positions. He was awarded the Distinguished Service Cross, the second highest medal for gallantry after the Medal of Honor.

SAMUEL S COURSEN

MEDAL OF HONOR
1ST LT
US ARMY
AUG 4 1926
OCT 12 1950
CLASS OF 1949
USMA

Samuel Coursen was awarded the Medal of Honor fighting in Korea on October 12, 1950. He died sixteen months after graduation.

Cadet Coursen's portrait from the 1949 yearbook. (Special Collections, West Point Library)

Samuel Coursen, First Lieutenant, (Class of 1949) was killed in action in Korea on October 12, 1950. He was awarded the Medal of Honor for conspicuous gallantry and intrepidity above and beyond the call of duty as an infantry platoon leader while attacking the enemy.

Edward White, Lieutenant Colonel, U.S. Air Force (Class of 1952), died in a fire on January 27,

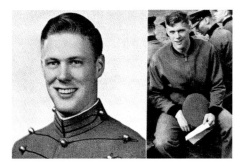

Edward White died in a fire on January 27, 1967, while training on the launch pad in the Apollo I command module.

A 1952 yearbook photograph of Edward White. (Special Collections, West Point Library)

1967, while conducting a training mission on the launch pad in the Apollo I command module with two other astronauts. They were the country's first astronaut fatalities aboard a spacecraft. This tragic accident caused a redesign of the module and hatch. Ed White is the only West Point graduate to have died as part of America's manned space program.

Andre Lucas, Lieutenant Colonel, (Class of 1954) was killed in action in Vietnam on July 23, 1970. For his extraordinary heroism while defending a fire support base as an infantry battalion commander, he was awarded the Medal of Honor.

Laura Walker, First Lieutenant (Class of 2003), was killed in action in Afghanistan on August 18, 2005. An engineer officer, Walker provided construction support to remote outposts, built an airfield, and worked on a paved road for the Afghan people. She was on her second combat tour,

serving in Iraq prior to Afghanistan. Laura Walker is the first female West Point graduate killed in action.

The spirit of West Point is most strongly felt when visiting the cemetery. A passage from the song *The Corps*—"That we of the Corps are treading where they of the Corps have trod"—best describes the feeling experienced while walking among the headstones of the fallen. Upon exiting the cemetery, a stanza from the *Alma Mater*— "Well done, be thou at peace"—comes forth to the memory of graduates and assumes greater meaning with the passage of time from that first Pass in Review as the Long Gray Line stretches ever onward. Visitors feel heartfelt gratitude for the service and sacrifice represented among the gallery of the fallen.

ABOVE LEFT: Cadet Lucas from the 1954 yearbook. (Special Collections, West Point Library)
ABOVE RIGHT: Cadet Laura Walker's portrait from the Class of 2003. (Special Collections, West Point Library)

APPENDIX A

Authorized Uniforms of the Present Day

Cadet uniforms of the present and the policy on wearing uniforms are similar to those of the past. Thayer banned cadet wear of civilian clothes in September 1821 to foster discipline among the cadets. Plebes are still not authorized to wear civilian clothes except when on leave. Upperclass cadets may wear the cadet casual uniform while First Class cadets may wear civilian clothes.

Most of the uniforms are made at the Cadet Uniform Factory located at West Point. This facility is unique as the only maker of uniforms among the five service academies. The other four academies— Naval, Air Force, Coast Guard, and Merchant Marine—contract out for their uniforms. The CUF also is unique in being a self-supporting, government-run operation. Cadets pay for their uniforms, as they have since 1802, out of their monthly stipend. The cost of a uniform pays for the material and labor of the dedicated government employees of the CUF. The CUF uses 275 different material items to manufacture uniforms along with items required for parade such as red sashes, shoulder belts, sword belts, waist belts shoulder boards, and guidons.

Three sections at the CUF make and repair cadet uniforms. The cutting and development section cuts the cloth into patterns for the appropriate garment and prepares the item for sewing. The tailoring section sews, alters, and repairs the iconic full dress, dress gray, and long overcoats. The machine operators group makes trousers, skirts, India white coats, parkas, and bathrobes.

The Cadet Color Guard marches on the Plain wearing gray over white.

A new roll of black reprocessed wool is being cut into sections to make the parka by the cutting room.

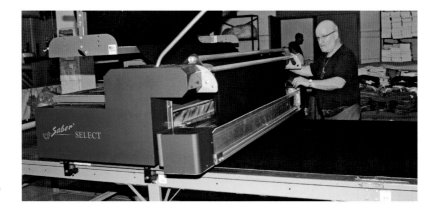

BOTTOM LEFT: Jae Yu, Class of 2011, wears the full dress over white under arms with white service cap, which is worn only for Graduation Day.

BOTTOM RIGHT: A skilled seamstress hand sews part of a full dress coat in the tailoring section.

"Cadet Gray Elastique" is the name of the 100-percent worsted wool cloth used to make most gray uniforms. Approximately 17,000 yards of this cloth, costing about $32 per yard, is used per year. Garments made from this cloth are the full dress coats, dress gray coats, dress trousers, and long overcoats. A full dress coat averages 1.5 yards of cloth

FAR LEFT: Stephanie Whitaker (left) wears dress mess, and Maria Casaletto (right) wears full dress with optional white skirt (both Class of 2011).

LEFT: India white worn by Thomas Kendall, Class of 2011.

while a long overcoat requires 3.75 yards per average coat. A dress gray coat requires, on average, 1.5 yards depending on the cadet's size.

Mock Elastique is a lighter weight (12.5 oz) blend of wool and polyester made to mimic the look of the all-wool cloth. The ME gray trousers are worn with the class shirt. In the summer during Beast Barracks, the trousers are worn with a short-sleeved open-collar white shirt called the white over gray uniform. ME skirts are made for female cadets.

India white is a 50/50 cotton and polyester blend cloth used for the India white coats, trousers, and skirts. These items are worn in spring, summer, and early fall.

Reprocessed wool is used to produce the black parka. Class year patches are sewn on the left front of the parka (see Appendix C). This is worn in the winter.

The cadet bathrobe is made of cotton velour fleece. The material is died to match the color of the traditional gray bathrobe made of wool.

Uniforms of the West Point Band also are manufactured by the CUF. The blue dress collared coat, dress blue trousers, summer white trousers, and regular and evening-length skirts are bought by the West Point Band for their musicians.

The CUF makes uniforms based on the findings of the Cadet Uniform Board, which meets twice a year to make recommendations to the superintendent on uniform changes. The CUB is composed of officers representing the Commandant of Cadets, the manager of the CUF, cadet services division chief, and a cadet from each class. Online surveys of cadets provide feedback on uniform item issues. A trend over the past few years is to make more items similar to Army uniforms. The cadet gray jacket was replaced by the Army black jacket starting with the Class of 2010. The cadet short overcoat was discontinued after the Class of 2011.

The cadet uniform regulation divides uniforms into five categories: Formal, Class, Athletic, Casual, and Accessories. A variety of combinations are worn throughout the year.

TOP LEFT: Dress gray worn by Nathan Wilson, Class of 2012.

TOP RIGHT: The female blouse (right) is optional wear with the short-sleeved class uniform worn all year.

The oldest formal uniform is full dress gray. There are two versions: one wearing gray trousers for winter, and the most familiar to the public called full dress gray over white. This uniform is worn at graduation and for most parades, with white trousers. Women may wear an optional white skirt with full dress for social occasions. For parades the full dress hat with plume is worn. The gray service cap is worn for other events. At graduation, the graduating class wears the white service cap.

India white is the summertime equivalent of full dress and is only seen by the public when the new cadets march out onto the Plain for their oath ceremony and the cadre wears the all white. This uniform

FAR LEFT: Michael Grdina, Class of 2012, wears the Army Combat Uniform (ACU) with the field gear, called Modular Lightweight Load-carrying Equipment (MOLLE) with the Army combat helmet

LEFT: Richard Naseer, Class of 2011, wears the ACU with patrol cap. His rank of cadet lieutenant is shown by the three black bars worn centered below the neck..

From left, William Balogh wears the Gym A with running suit, Samuel Preston wears G-3 Company athletic uniform for area hockey, William Osilaja wears Gym A with long-sleeve T-shirt, and Francis Park wears Gym A with short-sleeve T-shirt. All are Class of 2011.

also is worn for Ring Weekend and during Graduation week. The white service cap is worn with this uniform.

"Dress mess" is an optional uniform for women that they may wear to evening social occasions in lieu of full dress gray. A long black skirt is worn with this uniform. No headgear is worn with this uniform.

Dress gray is a semiformal uniform and also offers two versions with either gray or white trousers. The gray service cap is worn with this uniform. The least formal uniform is white over gray, consisting of a short-sleeved, open-collar white shirt worn over the ME trousers. Originally worn only during the summer, this uniform is now also authorized for spring and fall wear as the standard dress uniform. Optional for women are a gray skirt and white blouse. The white service cap is worn with this uniform.

RIGHT: Trevyn Hubbs, Class of 2012, in long overcoat with gray service cap and black gloves.

FAR RIGHT: Casual Uniform worn by Adam Akridge, Class of 2012.

FAR LEFT: John Williams, Class of 2012, wears the parka with knit cap.

LEFT: Black jacket with G-3 company patch on right shoulder with garrison cap worn by Kevin Zhang, Class of 2012.

Daily wear during the academic year consists of the "as for class" uniform. Cadets now wear the short-sleeved, open-collar shirt all year with gray trousers. A long-sleeved shirt with tie is optional though seldom worn. Women may wear a blouse and gray skirt. The garrison cap is worn with this uniform.

Field training during the summer is conducted wearing the Army Combat Uniform. This uniform is also worn during the academic year for special events such as the week prior to the Army–Navy game instead of the class uniform. This is an Army uniform the cadets may use after graduation. Brown combat boots and the ACU cap also are worn with this uniform. Field gear, called Modular Lightweight Load-carrying Equipment is issued for training at West Point and turned in prior to graduation. The Army combat helmet, made of Kevlar, is also worn for training.

The standard athletic uniform is the "Gym A." Black shorts are worn with a gray short or long-sleeved T-shirt with a cadet's name printed above the USMA crest. White socks and running shoes complete this basic athletic uniform. A running suit with charcoal gray jacket and black pants may be worn over the Gym A. Another option for wear with Gym A is a gray sweatsuit with gray sweatpants and USMA gray sweatshirt. Numerous athletic teams, club teams, and company intramural teams have their own prescribed uniform only authorized for use while participating in sport activities.

A casual uniform, consisting of a polo shirt and long khaki cotton pants, is authorized for Third Class cadets and above. Cadets must buy a shirt with an approved logo. Brown or black leather shoes

TOP LEFT: Stephen Roy, Class of 2011, dons the bathrobe to wear from his room to the showers and restrooms located in the center of the barracks.

TOP RIGHT: Kyle Tuttle, Class of 2011, wears raincoat with rain cover over the gray service cap.

along with a brown or black belt complete this optional uniform. Women may buy a khaki skirt.

The long overcoat is the oldest and most widely seen outer garment worn by cadets. Every year, before a television audience of millions, the cadets march onto the field for the Army–Navy game wearing their long overcoats. Black gloves are always worn with this item. The coat is authorized to be worn with full dress, dress gray, and white over gray.

During the fall, winter, and spring, the black jacket is worn with the class uniform. The cadet parka also may be worn with the class uniform during very cold winter days. The parka is also authorized to be worn with athletic uniforms, casual uniforms, and civilian clothes during the winter. The knit cap is worn with the parka along with black leather gloves.

The last outerwear item is the gray raincoat, which is similar to the long overcoat with a cape. The gray service cap with a rain cover worn over it is worn with the raincoat. If the cadet is wearing white over gray, the rain cover is not worn with the white service cap.

Men and women cadets live on the same floors in the barracks. Women only room with women, and men with men. A bathrobe is required when using the showers and restrooms, separate facilities for each gender, located in the center areas on each floor. The present-day, all-cotton robe is based on the early twentieth century designed all-wool robe.

APPENDIX B

Chronology of Events

1794 May 9: Thirty-two cadets, the first in the U.S. Army, are authorized by Congress as part of the Corps of Artillerists and Engineers, with two cadets per company and sixteen companies forming the Corps.

1799 December 12: In a letter written just two days before his death, George Washington advocates for a military academy.

1801 July 1: Cadets of the Corps of Artillerists and Engineers are ordered by the secretary of war to report to West Point.

September 1: School for cadets opens at West Point with George Baron as instructor of mathematics for fourteen cadets.

1802 March 16: President Thomas Jefferson signs an act of Congress establishing a military academy at West Point as part of the Corps of Engineers. Up to ten cadets are authorized within the Engineers along with forty cadets in the Artillery. Jonathan Williams is appointed first superintendent.

July 4: Formal opening of the U.S. Military Academy at West Point, New York, with ten cadets, the superintendent, and two instructors.

October 12: Joseph Swift and Simon Levy are commissioned into the Corps of Engineers as the first two graduates of the Academy. Both began as cadets in 1801.

1810 April 30: The distinctive cadet blue uniform is authorized. Prior to this, cadets wore uniforms of either the Engineers or the Artillery.

1812   April 29: Congress passes an Act establishing the United States Corps of Cadets and authorizing 250 cadets. In addition, Engineers retain up to ten cadets. The Act specified the formation of cadet companies and a summer encampment to train cadets. The 1st and 2nd Cadet Companies were established.

1814   Summer: Academy Superintendent Alden Partridge introduces gray for the cadet uniform.

1815   Summer: Partridge designs the full dress uniform with leather caps and white cross belts as full dress.

1816   September 4: The gray uniform is approved by the secretary of war, but with the civilian round hat instead of the leather shako. Gray cloth was less expensive than blue cloth, though a popular legend persists that gray was selected to honor the victory of the gray-clad soldiers of General Winfield Scott over the British at the Battle of Chippewa in 1814.

1817   July 28: Sylvanus Thayer relieves Partridge as superintendent. During the next sixteen years, Thayer will transform the Academy into the best engineering school in the world.

        September 15: Thayer appoints George Gardiner to supervise cadets in a position later titled the Commandant of Cadets.

        September 23: Chevrons are introduced as cadet rank, a modified version still used today.

1818   Summer: First summer encampment is held on the west side of the Plain.

        Fall: The bell-crowned leather shako is introduced. A smaller, lighter version of the shako is worn to this day on parade.

1819   Summer: Summer encampment is conducted on the northeast part of the Plain near Fort Clinton. Camps continue at this location until 1942.

1821   August 11: First colors are presented to the Corps of Cadets by the City of Boston.

1827   April 3: Companies are increased to four, numbered 1st through 4th, to increase leadership opportunities for the cadets.

1828   November 6: The single-breasted long overcoat of gray wool is first worn by cadets.

1833   July 1: Thayer resigns of his own accord after serving the longest (sixteen years) of any superintendent of the Academy to date.

1843 March 1: Congress approves a law for one cadet appointment per Congressional district to provide geographic diversity to the Corps of Cadets.

1848 March 27: Cadets on leave are authorized to wear blue frock coats.

1849 October 22: Riding jacket and pants are first issued.

1851 June 12: The overcoat is redesigned as double-breasted, a version worn to this day.

1870 June: White linen jacket and trousers are issued for summer camp.

1873 March 4: Cadets march for the first time in the President's inaugural parade—and in all subsequent inaugurations.

1877 June 14: The first African-American cadet, Henry Flipper, graduates.

1889 June 15: Gray fatigue coat, now called dress gray, replaces gray riding jacket.

1890 November 29: The first Army–Navy game is played on the Plain at West Point.

1898 Summer: Gray field service uniform along with field equipment replaces white linen fatigues.

October 13: The Coat of Arms and the motto "Duty, Honor, Country" are adopted.

1899 March 24: Black, gray, and gold are chosen as the colors for athletic uniforms.

April 26: New full dress hat, similar to first one of 1815, and white cross belts, also from 1815, are authorized and are first worn on July 3, 1899.

1900 August 24: Companies E and F are formed to accommodate a Corps of 481 authorized cadets organized as companies A–F.

1902 June 9–12: Centennial of the Academy is observed with President Theodore Roosevelt as the featured guest.

1913 December 15: All-white dress uniform is authorized, now called India white.

1917 June 8: Companies G, H, and I are added to Corps for nine total companies with 1,336 cadets.

1919 July 19: Three more companies, K, L, and M are formed.

1935 Size of Corps is increased to 1,964, but there remain just twelve companies.

1942 August: World War II prompts the size of Corps to increase to 2,500, organized into two regiments of eight companies each.

1946 Fall: The number of companies is increased to twelve per regiment or twenty-four total to increase leadership opportunities for cadets. Gray jacket is approved for wear.

1952 September 1: Black wool parka is authorized for winter wear.

1964 February: Congress approves expansion of Corps to 4,400 over a five-year period.

1965 June 30: Four regiments are formed with six companies per regiment, gradually increased to nine companies per regiment for thirty-six companies by 1969.

1976 July 7: 119 women are admitted to West Point.

1980 May 28: The Class of 1980 graduates, including the first sixty-two women graduates.

1982 September: Academic Board approves majors starting with the Class of 1985

1993 Fall: Congress reduces the Corps of Cadets to 4,000.

1998 Summer: Four I companies, one per regiment, are deactivated, with the Corps now composed of eight companies per regiment.

2003 Fall: Strength of Corps again increased to 4,400.

2006 Fall: The entering Class of 2010 is issued a black jacket in place of the gray version.

2011 May 20: Four I companies, I-1, I-2, I-3, and I-4, are reactivated.

APPENDIX C

Company Patches for Academic Year 2010–11

A-1 Axemen

B-1 Barbarians

C-1 Crusaders

D-1 Ducks

E-1 Vikings

F-1 Firehouse

G-1 Greeks

H-1 Root Hawgs

A-2 Spartans

B-2 Bulldogs

C-2 Circus

D-2 Dragons

E-2 Dawgs

F-2 Zoo

G-2 Gators

H-2 Hornets

A-3 Anacondas

B-3 Bandits

C-3 Game Cocks

D-3 Devils

E-3 Eagles

F-3 F-Troop

G-3 Gophers

H-3 Hurricanes

A-4 Apaches

B-4 Buffalos

C-4 Cowboys

D-4 Dukes

E-4 Elvis Lives

F-4 Frogs

G-4 Guppies

H-4 Hogs

INDEX

References to illustrations appear in **bold**.